SpringerBriefs in Applied Sciences and Technology

More information about this series at http://www.springer.com/series/8884

Ludovic Chamoin · Pedro Díez
Editors

Verifying Calculations – Forty Years On

An Overview of Classical Verification Techniques for FEM Simulations

Springer

Editors
Ludovic Chamoin
ENS Cachan
Cachan Cedex
France

Pedro Díez
UPC Barcelona Tech
Barcelona
Spain

ISSN 2191-530X ISSN 2191-5318 (electronic)
SpringerBriefs in Applied Sciences and Technology
ISBN 978-3-319-20552-6 ISBN 978-3-319-20553-3 (eBook)
DOI 10.1007/978-3-319-20553-3

Library of Congress Control Number: 2015943353

Springer Cham Heidelberg New York Dordrecht London

Printed on acid-free paper

Springer International Publishing AG Switzerland is part of Springer Science+Business Media (www.springer.com)

Preface

Numerical modeling and simulation is increasingly used as a complement to experimental modeling and analysis and as a design tool in engineering applications. Each of these numerical solutions is intrinsically carrying an error associated with the discretization (mesh) the modeler has decided to use. This decision is based on finding a tradeoff between the computational cost and the numerical quality. However, after almost forty years of worldwide and active research efforts, the problem of properly assessing and controlling the quality of the numerical simulations is still relevant and an issue of major interest. Currently, certain maturity has been reached and calculations for industrial applications can be verified and error bounds can be provided for many cases (even though they are rarely computed in practice). However, the design of sophisticated engineering systems requires increasingly complex and coupled modeling for which verification tools are missing. Furthermore, new issues are appearing as industry needs faster calculations for real-time decision making, design optimization, inverse analysis, or simulation-based control purposes, which urgently requires new strategies for mastering and certifying calculations, bounding errors, in particular in presence of uncertainties.

The present textbook is edited as a companion support of a pre-conference course given on the occasion of the ADMOS 2015 conference, held in Nantes (France) during June 7–10, 2015. It aims at providing the bases to error assessment tools, including state-of-the-art achievements on a posteriori verification in scientific computing. These topics pertain to the field of estimation of discretization errors associated with Finite Element simulations, with a focus on Computational Mechanics applications. This research discipline effectively enables to control the accuracy of numerical simulations and to drive adaptive algorithms. The document also aims at presenting recent advances and forthcoming research challenges on the subject. The content is made of four chapters written by expert researchers on the field, which present fundamental principles on classical a posteriori verification methods: explicit residuals methods, implicit residual methods, smoothing (recovery) methods, and duality-based methods.

We expect this book will help the reader to acquire an overview and insights into classical and state-of-the art techniques and tools for numerical verification.

Ludovic Chamoin
Pedro Díez

Contents

Explicit Residual Methods

Yvon Maday

Abstract Numerical methods are now well established for the approximation of the solution to partial differential equations. These simulations allow to better understand complex phenomenon and lead to their control and optimization. The accuracy of the solution needs nevertheless be certified and in some case improved and the measurement of the error between the exact solution to the problem and the approximated one provided by the computer simulation must be estimated. A large amount of research has been done in this direction. This paper summarizes some of the most classical approaches that are available and allow, at very little extra computational cost to certify the results. These are known as the explicit residual techniques.

Keywords A posteriori error estimation · Explicit residual techniques · Numerical convergence · Finite element method · Reduced basis approximation

1 Introduction

The numerical simulation of mathematical models, written under the form of a partial differential equation, performed on computers is most of the time centered on matrix inversion. Indeed, the numerical solution is sought in a finite dimensional space of functions provided with an appropriate basis set; the model is then transformed under the form of a system of possibly nonlinear equations involving the components of the numerical solution in the basis set. The choices of

- the finite dimensional space—denoted in what follows as X_N and is assumed to be of dimension $d(N)$
- the appropriate basis set of X_N

Y. Maday (✉)
Laboratoire Jacques-Louis Lions, CNRS, Institut Universitaire de France,
Sorbonne Universités, UPMC Univ Paris 06, Univ. Paris Diderot,
Sorbonne Paris Cité, F-75005 Paris, France
e-mail: maday@ann.jussieu.fr

Y. Maday
Division of Applied Mathematics, Brown University, Providence, USA

L. Chamoin and P. Díez (eds.), *Verifying Calculations – Forty Years On*,
SpringerBriefs in Applied Sciences and Technology,
DOI 10.1007/978-3-319-20553-3_1

- the solution algorithms adequate for solving the linear and nonlinear equations associated with a suitable stopping criteria

determine at the end of the computation of a given problem the accuracy of the approximation. In addition, (i) on one hand, the mathematical analysis of the nature of the various models proposed in different fields has led to a classification of problems and a know how on the best way to approximate the solutions (elliptic, mixed, parabolic, hyperbolic problems, reaction, advection, diffusion dominant behavior, conservations laws…), (ii) on the other hand various discretization spaces, discretization approaches and algorithms have been proposed and the numerical analysis that has been performed on these various approximation methods has identified their properties and the requirements on the solution and the model that allow to get the full benefit from the discretization approaches. This allows, when one faces a new problem, to guide the construction of the numerical approximation and the understanding of what can be expected.

In addition to these a priori statements, when the approximated solution to a given model is computed, it is important to be able to measure the quality of what has been performed and also give indications on what should be done in order to improve the precision of the approximation. This is the aim of the a posteriori analysis. It allows to quantify, in some cases very accurately, the error between the exact solution to the given model and the approximated one, we refer to [1–3, 23] for more advanced monographs on the subject.

In opposition to what happens for approximation through interpolation for instance where at least some values of the exact solution are accessible explicitly, the definition of the solution through a model is only implicit, e.g. given under the abstract form: find u in a normed functional space—denoted in what follows as X—such that the functional equation $F(u) = 0$. In such a frame, the evaluation of the error between the (unknown) exact solution u and its approximation u_N in X_N, expressed like $u - u_N$, measured in some norm, is itself a problem. In the case where the problem is linear, the functional equation is written as $F(u) = Au - f$ where f is a given functional data and A is a linear operator. In order to be slightly more precise, we can assume that A is a continuous operator from X into a functional space Y, and f is assumed to be given in Y. In such a case, formerly at least, the solution u is given by $u = A^{-1}f$ hence

$$u - u_N = A^{-1}f - u_N = A^{-1}[f - Au_N] = -A^{-1}F(u_N) \qquad (1)$$

hence a way to measure $u - u_N$ is to consider the expression $F(u_N) = Au_N - f$ that is defined to be the *residual* of the equation evaluated on the approximated solution u_N.

In what follows, we shall present some variations upon this subject dealing with the proper link between the error between u and u_N measured in some norm and the computation of the residual and its evaluation in some other norm. We refer to the chapter Residual type error estimators of A. Huerta and P. Diez where Implicit residual techniques are presented.

2 Fourier Approximation

Let us start with the simple setting of a problem provided with periodic boundary conditions, i.e. problems set over $\mathbb{R}^d$, $d = 1, 2$, or 3 where the functions we are considering are assumed to be 2π periodic in each direction. The problem can thus be restricted to a unit cell $\Omega = (0, 2\pi)^d$ of a periodic lattice $\mathcal{R}$ of $\mathbb{R}^d$.

The natural functional spaces in this frame are the periodic Lebesgue spaces $L^p_\sharp(\Omega)$ (resp. periodic Sobolev spaces $W^{m,p}_\sharp(\Omega)$, $m \in \mathbb{N}$), with $p \in \overline{\mathbb{R}}$, $1 \leq p \leq \infty$, defined as being the restriction to Ω of 2π-periodic functions in $L^p_{loc}(\mathbb{R}^d)$ (resp. of functions in $W^{m,p}_{loc}(\mathbb{R}^d)$) and we note $H^m_\sharp(\Omega)$ the spaces $W^{m,p}_\sharp(\Omega)$ when $p = 2$.

The natural discretization in this periodic settings consists in using a Fourier basis. We denote by $(X_N)_{N>0}$ the family of finite-dimensional subspaces of X defined by

$$X_N = \text{Span}\left\{e_{\boldsymbol{k}} : \boldsymbol{x} \mapsto e^{i\boldsymbol{k}.\boldsymbol{x}}, |k_p| \leq N, p = 1, \ldots, d, \boldsymbol{k} = (k_1, \ldots, k_d) \in \mathbb{Z}^d\right\}.$$

Remind now that, for any $v \in L^2_\sharp(\Omega)$,

$$v(\boldsymbol{x}) = \sum_{\boldsymbol{k}\in\mathbb{Z}} \widehat{v}_{\boldsymbol{k}} e_{\boldsymbol{k}}(\boldsymbol{x}),$$

where $\widehat{v}_{\boldsymbol{k}}$ is the $\boldsymbol{k}$th Fourier coefficient of v:

$$\widehat{v}_{\boldsymbol{k}} := \int_\Omega v(\boldsymbol{x})\, \overline{e_{\boldsymbol{k}}(\boldsymbol{x})}\, d\boldsymbol{x} = \int_\Omega v(\boldsymbol{x})\, e^{-i\boldsymbol{k}.\boldsymbol{x}}\, d\boldsymbol{x}.$$

For any integer m, we now endow the Sobolev space $H^m_\#(\Omega)$ with the equivalent norm expressed in Fourier modes as follows

$$\|v\|_{H^m} = \left(\sum_{\boldsymbol{k}\in\mathbb{Z}^d} \left(1 + |\boldsymbol{k}|^2\right)^m |\widehat{v}_{\boldsymbol{k}}|^2\right)^{1/2}, \tag{2}$$

and we note that the definition of this scale of spaces can be actually extended over real indices s (the spaces $H^s_\#(\Omega)$ associated with negative values of s correspond to the dual spaces of $H^{-s}_\#(\Omega)$). In what follows we shall use only this definition of the H^s–norm (2). We obtain that for any $r \in \mathbb{R}$, and all $v \in H^r_\#(\Omega)$, the best approximation of v in $H^s_\#(\Omega)$ for any $s \leq r$ is

$$\Pi_N v = \sum_{\boldsymbol{k}\in\mathbb{Z}, |\boldsymbol{k}|\leq N} \widehat{v}_{\boldsymbol{k}} e_{\boldsymbol{k}}. \tag{3}$$

The more regular v (the regularity being measured in terms of the Sobolev norms H^r), the faster the convergence of this truncated series to v: for any real numbers r and s with $s \leq r$, we have, see e.g. [8]

$$\forall v \in H^r_\#(\Omega), \quad \|v - \Pi_N v\|_{H^s} \le \frac{1}{N^{r-s}} \|v\|_{H^r}. \tag{4}$$

2.1 A Simple Elliptic Problem

Let us consider the problem of finding a periodic solution u such that

$$-\Delta u + \mathscr{V} u = f \tag{5}$$

where the potential $\mathscr{V}$ and the right hand side f are given periodic functions (the potential $\mathscr{V}$ being assumed to satisfy $\mathscr{V} \ge 1$). We assume also that these data are regular enough so as to ensure that the solution u to this problem belongs to some $H^{s^*}_\sharp(\Omega)$ with s^* large enough. The natural space X to set this problem is $X = H^1_\sharp(\Omega)$, and the problem (5) is, in this frame, equivalent to the variational problem: find $u \in X$ such that

$$\forall v \in X, \quad \int_\Omega \nabla u(\boldsymbol{x}).\nabla v(\boldsymbol{x}) d\boldsymbol{x} + \int_\Omega \mathscr{V} u(\boldsymbol{x}) v(\boldsymbol{x}) d\boldsymbol{x} = \int_\Omega f(\boldsymbol{x}) v(\boldsymbol{x}) d\boldsymbol{x} \tag{6}$$

Here, with respect to the notations introduced in the introduction, the operator $A = -\Delta + \mathscr{V}$. The existence of the solution to this problem results from the ellipticity of A, indeed, thanks to the hypothesis $\mathscr{V} \ge 1$ done on the potential we have

$$\forall \varphi \in X, \quad \langle A\varphi, \varphi \rangle \ge \|\varphi\|^2_X \tag{7}$$

that proves that A is an isomorphism from X into its dual space $X' = H^{-1}_\sharp(\Omega)$, in addition the inverse A^{-1} is continuous with norm 1.

The classical approximation for this problem is obtained by the Galerkin settings where the variational formulation is degraded over the space X_N: find $u_N \in X_N$ such that

$$\forall v_N \in X_N, \quad \int_\Omega \nabla u_N(\boldsymbol{x}).\nabla v_N(\boldsymbol{x}) d\boldsymbol{x} + \int_\Omega \mathscr{V}(\boldsymbol{x}) u_N(\boldsymbol{x}) v_N(\boldsymbol{x}) d\boldsymbol{x} = \int_\Omega f(\boldsymbol{x}) v_N(\boldsymbol{x}) d\boldsymbol{x} \tag{8}$$

Note that an additional use of Gauss integration rule may be (and is generally) required to compute the potential part, we shall not dwell upon this here.

The classical Cea lemma states that a solution u_N exists for every N and

$$\|u - u_N\|_{H^1} \le C \min_{v_N \in X_N} \|u - v_N\|_{H^1} \tag{9}$$

and in particular since the solution u is regular enough, we deduce from (4) that

$$\|u - u_N\|_{H^1} \le C N^{1-s^*} \tag{10}$$

This is a priori analysis that does not allow to measure exactly the error one the computation has been performed since the constant C that appears in this bound is not easy to derive and s^* is not known either.

2.2 A Posteriori Analysis in the Periodic Setting

We shall thus consider the approach sketched in the introduction and proceed from (1):

$$\|u-u_N\|_X = \|A^{-1}[f-Au_N]\|_X \leq \|f-Au_N\|_{X'} = \|f-(\Delta u_N+\mathscr{V} u_N)\|_{X'} \quad (11)$$

Once the approximate solution u_N has been computed, the evaluation of the dual norm X' is achieved from the knowledge of the Fourier coefficients of f and $\mathscr{V}$ from the formula (2). It is noteworthy to realize that (8) is equivalent to

$$\pi_N[f-(\Delta u_N+\mathscr{V} u_N)] = 0 \quad (12)$$

stating, in other words that the Fourier coefficients of $f-(\Delta u_N+\mathscr{V} u_N)$ on indexes with $|k_p| \leq N$ are zero. It then follows easily that

$$\|f-(\Delta u_N+\mathscr{V} u_N)\|_{X'} \leq \frac{1}{N}\|f-(\Delta u_N+\mathscr{V} u_N)\|_{L^2} \quad (13)$$

which makes the computation of the a posteriori estimator more easy since this can be evaluated as an integral through, e.g. the use of a quadrature rule.

2.3 Stopping Criteria for Iterative Solution Algorithms

Even is the context of this simple problem is not so appropriate, because a very accurate value of u_N can indeed be computed, let us assume that the inversion of A in (8) is done iteratively, either through a Jacobi iteration of a (preconditioned) conjugate gradient method. It is interesting to proceed a little further than in (12) and introduce the operator $A_N = \pi_N \circ A \circ \pi_N$. Indeed, this operator is an isomorphism in X_N and (8) reads: find $u_N \in X_N$ such that

$$A_N u_N = \pi_N f. \quad (14)$$

We thus assume that this problem is solved through some iterative technique and u_N is contained as a limit, when m goes to infinity, of u_N^m. The point we want to address here is that the iterative procedure is never pushed to extreme convergence and thus

u_N is approximated by a u_N^M, where M is the maximum number of iterations that is accepted or the increment between u_N^M and u_N^{M-1} is below some threshold. Then, the solutions given by the code does not satisfy (14) exactly. The error at this level of iteration is thus decomposed as

$$\|u - u_N^M\|_X \leq \|f - (\Delta u_N^M + \mathcal{V} u_N^M)\|_{X'} = \|f - Au_N^M\|_{X'}$$

this error actually involves of two orthogonal contributions: $\pi_N[f - Au_N^M)]$ and $(Id - \pi_N)[f - Au_N^M]$, hence

$$\|u - u_N^M\|_X \leq \sqrt{\|\pi_N f - A_N u_N^M\|_{X'}^2 + \|(Id - \pi_N)[f - Au_N^M]\|_{X'}^2} \qquad (15)$$

The two terms on the right hand side are of different nature, the first one deals with the iterative process, if M goes to ∞ it will disappear. On the contrary, the second one has nothing to do with the convergence of the iterative process, it reveals the error due to the discretization truncation, i.e. the choice of N.

Note that due to (12), estimate (11) can also be written as

$$\|u - u_N\|_X \leq \|(Id - \pi_N)[f - Au_N]\|_{X'} \qquad (16)$$

which makes the link with (15) in the approximately converged situation.

This very simple example is an instantiation of a more general recent direction of research in the a posteriori analysis that allows to identify, among discretization errors (linked to the choice of basis set and number of degrees of freedom) and the algorithm errors (linked to the stopping criteria of the solution procedure, either for linear or non linear fixed point strategy). This really allow to monitor the approximation process in order to balance the various element that count on the accuracy and achieve a given certified accuracy.

We refer to [3, 14, 16] for more details on this approach and case studies.

3 Finite Element Approximation

In this section, we shall develop a similar strategy as above but we shall see that, associated to the local character of the finite element basis (with respect to the global nature of the Fourier—or plane wave—approximation) we shall be able to get more from the a posteriori analysis. In addition to a global error control that allows to certify the computation, the analysis provides also an error localization that allows to improve the accuracy by refining (or better positioning) the degrees of freedom.

For the sake of completeness, let us remind a few notations about the finite element approximation for second order elliptic problem. The domain Ω where the partial differential equation is set is assumed to be polygonal or polyhedral; it is assume to be provided with a family of simplicial partition $\{\mathcal{T}_h\}_h$ (denoted also meshes) such that $\overline{\Omega} = \cup_{K\in\mathcal{T}_h} K$, where each K is

- in 2 dimensions a triangle
- in 3 dimensions a tetrahedron

The use of quadrilateral of hexahedron elements is also possible. The intersection of two such simplexes is either empty, a common vertex, a common entire edge or (in 3d) a common entire face.

We then assume that these simplexes are shape regular in the sense that the ratio of the diameter h_K of any simplex K in $\mathcal{T}_h$ on the diameter ρ_K of the largest ball inscribed in K is upper bounded by a constant independent of h.[1] We denote by $\boldsymbol{n}_K$ the outward unit normal vector to K.

The finite element space built upon each mesh $\mathcal{T}_h$ is defined by

$$X_h = \left\{v_h \in \mathscr{C}^0(\overline{\Omega});\, \forall K \in \mathcal{T}_h,\, v_{h|K} \in \mathscr{P}_k(K)\right\} \cap X \tag{17}$$

where, depending upon the fact that the problem is set with Dirichlet or Neuman boundary conditions, $X = H_0^1(\Omega)$ or $X = H^1(\Omega)$. In this definition, $\mathscr{P}_k(K)$ denotes the space of polynomials with total degree $\leq k$ over K.

It is well known that the best approximation in X of regular functions by elements of X_h satisfy e.g. (for any $p \in \mathbb{R}$, $p \geq 1$)

$$\inf_{v_h\in X_h} \|v - v_h\|_{W^{s,p}(\Omega)} \leq Ch^{r-s}\|v\|_{W^{r,p}(\Omega)} \tag{18}$$

for any $s, 0 \leq s \leq 1, r, 1 \leq r \leq k+1$.

In this frame, the toy example of the Poisson equation: find $u \in X = H_0^1(\Omega)$ such that

$$\forall v \in X; \quad \int_\Omega \nabla u(\boldsymbol{x})\nabla v(\boldsymbol{x})d\boldsymbol{x} = \int_\Omega f(\boldsymbol{x})v(\boldsymbol{x})d\boldsymbol{x} \tag{19}$$

is approximated by u_h solution in X_h to the problem

$$\forall v_h \in X_h; \quad \int_\Omega \nabla u_h(\boldsymbol{x})\nabla v_h(\boldsymbol{x})d\boldsymbol{x} = \int_\Omega f(\boldsymbol{x})v_h(\boldsymbol{x})d\boldsymbol{x} \tag{20}$$

Here again, there exists a unique solution u_h, for any mesh $\mathcal{T}_h$, and Cea's Lemma states that, provided that the solution $u \in H^r(\Omega)$, with $1 \leq r \leq k+1$,

$$\|u - u_h\|_1 \leq Ch^{r-1}\|u\|_{H^r(\Omega)} \tag{21}$$

with the same comment as in Sect. 2 that neither C nor $\|u\|_{H^r(\Omega)}$ are sufficiently known to let (21) provide any certification on the accuracy of the approximation.

[1] As is usual, h is a bague notion that does not identify uniquely $\mathcal{T}_h$ but represents the largest diameter h_K, for K in $\mathcal{T}_h$.

In order to achieve this goal, the residual is again central: following (1), we have

$$|u - u_h|_1 = \|f - \Delta u_h\|_{H^{-1}(\Omega)} \tag{22}$$

which is accurate but useless (even though this has an a posteriori nature) because the negative norm, contrarily to what happens in the Fourier discretization, is not directly computable.

3.1 Evaluation of the Residual

In order to proceed, we go back to the definition of the negative norm

$$\|f - \Delta u_h\|_{H^{-1}(\Omega)} = \sup_{v \in H_0^1(\Omega)} \frac{\langle f - \Delta u_h, v\rangle_{H^{-1},H_0^1}}{|v|_1} \tag{23}$$

where $\langle .,.\rangle_{H^{-1},H_0^1}$ stands for the duality product between $H^{-1}(\Omega)$ and $H_0^1(\Omega)$. Assuming some regularity on $f \in L^2(\Omega)$—consistent with the fact that u is more regular than $H^1(\Omega)$, at least $H^2(\Omega)$—we have, by definition of the weak derivative

$$\begin{aligned}\langle f - \Delta u_h, v\rangle_{H^{-1},H_0^1} &= \int_\Omega f(\boldsymbol{x})v(\boldsymbol{x})d\boldsymbol{x} - \int_\Omega \nabla u_h(\boldsymbol{x}).\nabla v(\boldsymbol{x})d\boldsymbol{x}\\ &= \int_\Omega f(\boldsymbol{x})(v(\boldsymbol{x}) - v_h(\boldsymbol{x}))d\boldsymbol{x} - \int_\Omega \nabla u_h(\boldsymbol{x}).\nabla(v(\boldsymbol{x}) - v_h(\boldsymbol{x}))d\boldsymbol{x}\end{aligned}$$

where the second line follows from (20), Using the fact that $\overline{\Omega} = \cup_{K\in\mathscr{T}_h} K$, we can decompose the above integral as follows

$$\langle f - \Delta u_h, v\rangle_{H^{-1},H_0^1} = \sum_{K\in\mathscr{T}_h} \left(\int_K f(\boldsymbol{x})(v(\boldsymbol{x}) - v_h(\boldsymbol{x}))d\boldsymbol{x} - \int_K \nabla u_h(\boldsymbol{x}).\nabla(v(\boldsymbol{x}) - v_h(\boldsymbol{x}))d\boldsymbol{x}\right) \tag{24}$$

A simple integration by parts over each of the simplexes $K \in \mathscr{T}_h$ can be performed, it makes some contribution on the union of each boundary ∂K, the union of which is denoted as $\mathscr{S}_h = \cup_{K\in\mathscr{T}_h} \partial K$. Note that $\mathscr{S}_h$ is composed of $d-1$–faces through each of them, an element $w_h \in X_h$ is continuous, but the gradient of which need not be continuous and thus, across each face $F \in \mathscr{S}_h$, between two adjacent elements K and K' of $\mathscr{T}_h$, there is a jump denoted as $[\frac{\partial v_h}{\partial n}]$ with

$$[\frac{\partial v_h}{\partial n}] = \nabla(v_{h|K}).\boldsymbol{n}_K + \nabla(v_{h|K'}).\boldsymbol{n}_{K'} \tag{25}$$

The result of the integrations by parts over each K applied to (24) leads to

$$
\begin{aligned}
\langle f - \Delta u_h, v\rangle_{H^{-1},H_0^1} &= \sum_{K\in\mathscr{T}_h} \left(\int_K (f(\boldsymbol{x}) + \Delta u_h(\boldsymbol{x}))\,(v(\boldsymbol{x}) - v_h(\boldsymbol{x}))d\boldsymbol{x} \right.\\
&\qquad \left. - \int_{\mathscr{S}} \left[\frac{\partial u_h}{\partial n}\right](s)(v(s) - v_h(s))ds \right)\\
&\le \sum_{K\in\mathscr{T}_h} \|f - \Delta u_h\|_{L^2(K)} \|v - v_h\|_{L^2(K)}\\
&\quad + \sum_{F\in\mathscr{S}_h} \|\left[\frac{\partial u_h}{\partial n}\right]\|_{L^2(F)} \|v - v_h\|_{L^2(F)}
\end{aligned}
$$

We now recall some results about the regularization operator π_h^C introduced by P. Clément in [9] and refer to e.g. [4] for the following properties: for any $v \in H^1(\Omega)$,

$$
\|v - \pi_h v\|_{L^2(K)} \le c h_K \|v\|_{H^1(\Delta_K)} \tag{26}
$$

$$
\|v - \pi_h v\|_{L^2(F)} \le c h_F^{\frac{1}{2}} \|v\|_{H^1(\Delta_F)} \tag{27}
$$

valid for any simplex K (resp. for any face F of K) with Δ_K (resp. Δ_F) being the union of simplexes K' that intersect K (resp. that intersect F). This leads to

$$
\langle f - \Delta u_h, v\rangle_{H^{-1},H_0^1} \le c \left(\sum_{K\in\mathscr{T}_h} \eta(K)^2 \right)^{\frac{1}{2}} \|v\|_{H^1(\Omega)} \tag{28}
$$

with

$$
\eta(K) = \left(h_K \|f + \delta u_h\|_{L^2(K)} + \frac{1}{2} \sum_{F\in\mathscr{S}(K)} h_F^{\frac{1}{2}} \|\left[\frac{\partial u_h}{\partial n}\right]\|_{L^2(F)} \right) \tag{29}
$$

hence this proves, from (22)

$$
|u - u_h|_1 \le c \left(\sum_{K\in\mathscr{T}_h} \eta(K)^2 \right)^{\frac{1}{2}} \tag{30}
$$

This a posteriori estimator is composed of local quantities that contribute to the global error. If some of them are bigger that others, this indicates that these simplexes should be refined to improve the accuracy. Going over the same arguments, similar estimators involving the $L^p(K)$–norm of $f + \delta u_h$ and the $L^q(F)$–norm of $\left[\frac{\partial u_h}{\partial n}\right]$ have been proposed in [5]. These indicators are useful in the sens they constitute indeed a sharp estimate of the error. This is more complex to estimate and we refer to [5] for the proof (see also [1, 22]) of the following proposition

Proposition 1 *The following estimate is valid for every $K \in \mathcal{T}_h$*

$$\eta(K) \leq c(|u - u_h|_{H^1(\omega_K)} + h_k \|f - f_{mh}\|_{L^2(\omega_K)} \tag{31}$$

where ω_K is the union of all simplexes K' that share a $d-1$–face with K and f_{mh} is any good approximation of f in $L^2(\Omega)$ (not continuous) locally in $\mathcal{P}_m(K)$ with $m \geq k-2$.

3.2 Extension to the Stokes Problem

Let us now consider the Stokes problem, in the same settings as previously. This consists in finding a pair of velocity/pressure: $(\boldsymbol{u}, p) \in (H_0^1(\Omega))^d \times L_0^2(\Omega)$ (with $L_0^2(\Omega)$ being the subspace of $L^2(\Omega)$–functions with zero average) such that

$$\begin{aligned} \forall \boldsymbol{v} \in (H_0^1(\Omega))^d, \quad \nu \int_\Omega \nabla \boldsymbol{u}(\boldsymbol{x}).\nabla \boldsymbol{v}(\boldsymbol{x}) d\boldsymbol{x} - \int_\Omega \operatorname{div} \boldsymbol{v}(\boldsymbol{x}) p(\boldsymbol{x}) d\boldsymbol{x} &= \langle \boldsymbol{f}, \boldsymbol{v} \rangle \\ \forall q \in L^2(\Omega), \quad \int_\Omega \operatorname{div} \boldsymbol{u}(\boldsymbol{x}) q(\boldsymbol{x}) d\boldsymbol{x} &= 0 \end{aligned} \tag{32}$$

ν here is a positive viscosity and $\boldsymbol{f}$ is given in $(L_0^2(\Omega))^d$.

The analysis of this problem, together with the numerical analysis of its discretization with finite element is classical and we refer to [12] and BF for details. An important element for the analysis is the *inf-sup* condition, both at the continuous level

$$\inf_{q \in L_0^2(\Omega)} \sup_{\boldsymbol{v} \in (H_0^1(\Omega))^d} \frac{\int_\Omega \operatorname{div} \boldsymbol{v}(\boldsymbol{x}) q(\boldsymbol{x}) d\boldsymbol{x}}{\|\boldsymbol{v}\|_{H^1} \|q\|_{L^2}} \geq \beta > 0 \tag{33}$$

and at the discrete level

$$\inf_{q_h \in M_h} \sup_{\boldsymbol{v}_h \in X_h} \frac{\int_\Omega \operatorname{div} \boldsymbol{v}_h(\boldsymbol{x}) q_h(\boldsymbol{x}) d\boldsymbol{x}}{\|\boldsymbol{v}_h\|_{H^1} \|q_h\|_{L^2}} \geq \beta_h > 0 \tag{34}$$

with, possibly, a constant β_h independent of h. Above, obviously, X_h is a finite element approximation associated to $(H_0^1(\Omega))^d$ (actually the dth product of the previous discrete space adapted to the Laplace problem will work) and M_h is a discrete space adapted to $L_0^2(\Omega)$ composed of (possibly discontinuous) piecewise polynomials.

The associated residual, analogous to the previous section is

$$\begin{aligned} \eta_S(T) = h_K \|\boldsymbol{f} + \nu \Delta \boldsymbol{u}_h - \nabla p_h\|_{L^2(K)^d} + \|\operatorname{div} \boldsymbol{u}_h\|_{L^2(K)} \\ + \frac{1}{2} \sum_{F \in \mathcal{S}(K)} h_F^{\frac{1}{2}} \| \left[\nu \frac{\partial \boldsymbol{u}_h}{\partial n} - p_h \boldsymbol{n} \right] \|_{L^2(F)^d} \end{aligned} \tag{35}$$

Similar lower bounds and upper bounds for the error in $\boldsymbol{u} - \boldsymbol{u}_h$ and $p - p_h$ can be obtained similarly as above and we refer to [5] for the the complete statements.

4 Reduced Basis Approximation

In this section we are going to present the above a posteriori technique applied to a more recent approach adapted to the approximation of the solutions of a class of problems depending on a parameter $\mu \in \mathcal{D}$ set in the form: find $u \equiv u(\mu) \in X$ such that $\mathcal{F}(u; \mu) = 0$. The set of parameters $\mathcal{D}$ is not too much specified here, it can either be a subset of $\mathbb{R}$, or $\mathbb{R}^p$, or even a subset of functions. Such problems arise in many situations such as e.g. optimization, control or parameter-identification problems, response surface or sensibility analysis. In case $\mathcal{F}$ is written through partial differential equations, the problem may be stationary or time dependent but in all these cases, a solution $u(\mu)$ has to be evaluated or computed for many instances of $\mu \in \mathcal{D}$.

4.1 Basics and Rational of the Reduced Basis Approach

In opposition to the previous Fourier or finite element approaches that are *multipurpose* in the sense that the general settings is adapted to the solution to any partial differential equation, the reduced basis method is only adapted to the general problem it has been built for.

These reduced basis approximations indeed consists in approximating the solution $u(\mu)$ of a parameter dependent problem $\mathcal{F}(u; \mu) = 0$ by a linear combination of appropriate, preliminary computed, solutions $u(\mu_i)$ for well chosen parameters μ_i, $i = 1, \ldots, N$. Note that accurate evaluations of these basis solutions need to be performed and this is done with more classical approximation methods in a off-line stage.

The rational of this approach, stands in the fact that the set $\mathcal{S}(\mathcal{D}) = \{u(\mu)$ of all solutions when $\mu \in \mu\}$ behaves well. In order to apprehend in which sense the good behavior of $\mathcal{S}(\mathcal{D})$ should be understood, it is helpful to introduce the notion of n-width following Kolmogorov [15] (see also [18]).

Definition 1 Let X be a normed linear space, A be a subset of X and X_n be a generic n-dimensional subspace of X. The deviation of A from X_n is

$$E(A; X_n) = \sup_{x \in A} \inf_{y \in X_n} \|x - y\|_X.$$

The *Kolmogorov n-width* of A in X is given by

$$\begin{aligned} d_n(A, X) &= \inf\{E(A; X_n) : X_n \text{ an } n\text{-dimensional subspace of X}\} \\ &= \inf_{X_n} \sup_{x \in A} \inf_{y \in X_n} \|x - y\|_X . \end{aligned} \tag{36}$$

The n-width of A thus measures the extent to which A may be approximated by a n-dimensional subspace of X.

There are many reasons why this n-width may go rapidly to zero as n goes to infinity. In our case, where $A = \mathscr{S}(\mathscr{D})$, we can refer to regularity of the solutions $u(\mu)$ with respect to the parameter μ, or even to analyticity [10]. An example is indeed provided by Kolmogorov stating that $d_n(\tilde{B}_2^{(r)}; L^2) = \mathscr{O}(n^{-r})$ where $\tilde{B}_2^{(r)}$ is the unit ball in the Sobolev space of all 2π-periodic real valued, $(r-1)$-times differentiable functions whose $(r-1)$st derivative is absolutely continuous and whose rth derivative belongs to L^2. Actually, exponential convergence is achieved when analyticity exists in the parameter dependancy. The knowledge of the rate of convergence is not sufficient: of theoretical interest is the determination of the (or at least one) optimal finite dimensional space X_n that realizes the infimum in d_n, provided it exists. For practical reasons, we want to restrict ourselves to finite dimensional spaces that are spanned by elements of $\mathscr{S}(\mathscr{D})$.

This is at the basics of the reduced basis method that proposes to choose properly a sequence of parameters $\mu_1, \ldots, \mu_n, \ldots \in \mathscr{D}$, then define the vectorial space $X_N = \text{Span}\{u(\mu_1), \ldots, u(\mu_N)\}$ and look for an approximation of $u(\mu)$ in X_N.

When applied e.g. to an elliptic problem: Find $u(\mu) \in X$ such that

$$a(u(\mu), v; \mu) = f(v), \quad \forall v \in X \,. \tag{37}$$

where X is some Hilbert space, and a is a continuous and elliptic, bilinear form in its two first arguments, regular in the parameter dependency and f is some given continuous linear form over X. We assume for the sake of simplicity that the ellipticity is uniform with respect to $\mu \in \mathscr{D}$: $\exists \alpha > 0$

$$\forall \mu \in \mathscr{D}, \forall u \in X, \quad a(u, u; \mu) \geq \alpha \|u\|_X^2 \,,$$

an that the continuity of a is uniform with respect to $\mu \in \mathscr{D}$ as well: $\exists \gamma > 0$

$$\forall \mu \in \mathscr{D}, \forall u, v \in X, \quad |a(u, v; \mu)| \leq \gamma \|u\|_X \|v\|_X \,.$$

It is classical to state that, under the previous hypothesis, problem (37) has a unique solution for any $\mu \in \mathscr{D}$. The Galerkin method is a standard way to approximate the solution to (37) provided that a finite dimensional subspace X_N on X is given. It consists in: Find $u_N(\mu) \in X_N$ such that

$$a(u_N(\mu), v_N; \mu) = f(v_N), \quad \forall v_N \in X_N \,, \tag{38}$$

which similarly has a unique solution $u_N(\mu)$. Cea's lemma then states that

$$\|u(\mu) - u_N(\mu)\|_X \leq (1 + \frac{\gamma}{\alpha}) \inf_{v_N \in X_N} \|u(\mu) - v_N\|_X \,. \tag{39}$$

The best choice for the basis element $u(\mu_1), \ldots, u(\mu_N)$ of X_N should be associated to finding the optimal choice associated with the definition of the Kolmogorov width. Unfortunately, there is no reason why these optimal spaces should be hierarchical and there does not even exist constructive way to determine these optimal spaces. We generally refer to a greedy algorithm such as the following one:

$$\begin{aligned} \mu_1 &= \arg \sup_{\mu \in \mathscr{D}} \|u(\mu)\|_X \,, \\ \mu_{i+1} &= \arg \sup_{\mu \in D} \|u(\mu) - P_i u(\mu)\|_X \end{aligned} \tag{40}$$

where P_i is the orthogonal projection onto $V_i = \text{span}\{u(\mu_1), \ldots, u(\mu_i)\}$ or a variant based on the evaluation of this projection error and relies on an a posteriori estimator (see next section). The convergence proof for the related greedy algorithm is somehow more complex and presented in a quite general settings in [6, 7, 11].

An a posteriori estimation can also be performed in this settings, using the same concept of the evaluation of the residual. More will be provided in the next subsection, we also refer to [17, 19, 20] and e.g. to [13] for parabolic problems.

4.2 A Posteriori Validation of Outputs

Most of the time though, the complete knowledge of the solution of the problem (37) is not required. What is required, is outputs computed from the solution $s = s(u)$, where s is some continuous functional defined over X. In order to have a hand over this output, the reduced basis method consists first in computing $u_N \in X_N$ solution of the Galerkin approximation (38), then propose $s_N = s(u_N)$ as an approximation of s. Assuming Lipschitz condition (ex. linear case) over s, it follows that

$$|s - s_N| \leq c\|u - u_N\|_X. \tag{41}$$

Thus any information over the error in the energy norm will allow to get verification (provided you are able to evaluate c). Actually it is well known that the convergence of s_N towards s most often goes faster. Indeed, let us assume we are in the linear output case where $s \equiv \ell$ is a linear continuous mapping over X. It is then standard to introduce the *adjoint state*, solution of the following problem: find $\psi \in X$

$$a(v, \psi; \mu) = -\ell(v), \quad \forall v \in X. \tag{42}$$

The error in the output is then (remember that, for any $\phi_N \in X_N$, $a(u, \phi_N; \mu) = a(u_N, \phi_N; \mu) = (f, \phi_N)$)

$$
\begin{aligned}
s_N - s &= \ell(u_N) - \ell(u) \\
&= a(u, \psi; \mu) - a(u_N, \psi; \mu) \\
&= a(u, \psi - \phi_N; \mu) - a(u_N, \psi - \phi_N; \mu), \quad \forall \phi_N \in X_N \\
&= a(u - u_N, \psi - \phi_N; \mu), \quad \forall \phi_N \in X_N \\
&\leq c\|u - u_N\|_X \|\psi - \phi_N\|_X, \quad \forall \phi_N \in X_N ,
\end{aligned} \tag{43}
$$

so that the best fit of ψ in X_N can be chosen in order to improve the first error bound (41) that was proposed for $|s - s_N|$.

For instance if ψ_N is the solution of the Galerkin approximation to ψ in X_N, we get

$$
|s - s_N| \leq c\|u - u_N\|_X \|\psi - \psi_N\|_X \, . \tag{44}
$$

Actually, the approximation of ψ in X_N may not be very accurate since X_N is well suited for approximating the elements $u(\mu)$ and—except in the case where $\ell = f$ named the compliant case—a separate reduced space $\tilde{X}_N$ should be built which provides an associated approximation $\tilde{\psi}_N$. Then an improved approximation for $\ell(u)$ is given by $\ell_{\text{imp}} = \ell(u_N) - a(u_N, \tilde{\psi}_N) + f(\tilde{\psi}_N)$ since (44) holds with $\|\psi - \tilde{\psi}_N\|_X$ for which a better convergence rate is generally observed.

Even improved, this result is still a priori business and it does not allow to qualify the approximation for a given computation. In order to get a posteriori information, between $\ell(u)$ and $\ell(u_N)$ (or ℓ_{imp}), we have to get a hand on the residuals in the approximations of the primal and dual problems. We introduce for any $v \in X$,

$$
\mathcal{R}^{pr}(v; \mu) = a(u_N, v; \mu) - \langle f, v\rangle, \quad \mathcal{R}^{du}(v; \mu) = -a(v, \tilde{\psi}_N; \mu) - \ell(v). \tag{45}
$$

We then compute the reconstructed errors associated with the previous residuals. These are the solutions of the following problems

$$
2\alpha \int \nabla \hat{e}^{pr(du)} \nabla v = \mathcal{R}^{pr(du)}(v; \mu), \quad \forall v \, , \tag{46}
$$

we then have

Theorem 1 *Let* $s_N^- = \ell_{\text{imp}} - \alpha \int [\nabla(\hat{e}^{pr} + \hat{e}^{du})]^2$ *then* $s^- \leq s$. *In addition, there exists two constants* $0 < c \leq C$ *such that*

$$
c|s - s_N| \leq s - s_N^- \leq C|s - s_N|.
$$

Proof We provide the proof since it is quite elementary. Let us denote by e_N the difference between the exact solution and the approximated one $e_N = u - u_N$. From (46), we observe that

$$
2\alpha \int \nabla \hat{e}^{pr} \nabla e_N = -a(e_N, e_N; \mu)
$$

and

$$2\alpha \int \nabla \hat{e}^{du} \nabla e_N = -a(e_N, \tilde{\psi}_N; \mu) - \ell(e_N) = f(\tilde{\psi}_N) - a(u_N, \tilde{\psi}_N) - \ell(e_N).$$

Taking this into account allows to write

$$\begin{aligned} \ell_{\text{imp}} - \alpha \int \nabla(\hat{e}^{pr} + \hat{e}^{du})^2 &= \ell(u_N) - a(u_N, \tilde{\psi}_N) + f(\tilde{\psi}_N) - \alpha \int \nabla(\hat{e}^{pr} + \hat{e}^{du})^2 \\ &= \ell(u) - \alpha \int \nabla(e_N + \hat{e}^{pr} + \hat{e}^{du})^2 - a(e_N, e_N; \mu) + \alpha \int [\nabla e_N]^2 , \end{aligned} \tag{47}$$

and the proof follows from the uniform ellipticity of $a(.,.;\mu)$.

Despite the fact that we have avoided to speak about any practical implementation so far, Theorem 1 is at this stage quite informative in the sense that in order to obtain s_N^-, the problem (46) to be solved, is parameter independent and simpler than the original one, provided that we have a good evaluation of the ellipticity constant. Note that an upper bound s_N^+ can also be obtained similarly as above. In Sect. 4.3 we shall explain how to transform these constructions in a method that can be implemented. Before this we should explain how the previous estimator may help in the greedy procedure presented above. It combines the reduced approximation and the error evaluation as follows:

- take a first parameter (randomly)
- use a (one dimensional) reduced basis approach over a set of parameter values (chosen randomly) and select, as a second parameter, the one for which the associated predicted error $s_1^+ - s_1^-$ is the largest.

this gives now a 2 dimensional reduced basis method.

- use this (2 dimensional) reduced basis approach over the same set of parameters and select, as a third parameter, the one for which the associated error is the largest.

this gives a 3 dimensional reduced basis method…

- and proceed…

Note that we then only compute accurately the solutions corresponding to the parameters that are selected this way.

The a posteriori approach that has been presented above relies on the uniform ellipticity of the bilinear form and the knowledge of the ellipticity constant. For more general problems, where only, nonuniform inf-sup conditions are valid (e.g. noncoercive Helmholtz acoustics problem which becomes singular as we approach resonance) smarter definitions should be considered. We refer to [21] for improved methods in this direction.

4.3 Black Box Implementation

The solution procedure involves the evaluation of the elements of the stiffness matrix $a(\zeta_i, \zeta_j; \mu)$, $1 \leq i, j \leq N$ that depends on the current parameter μ. This computation involves some derivatives and the evaluation of integrals, that have to be performed and this may be *very* lengthy. It should be stated here that the implementation of the reduced type method has to be much faster than the solution procedure that was used to compute the reduced basis, much means many order of magnitude. The $\mathcal{O}(\dim X_N)^2$ entrees of the stiffness matrix have thus to be evaluated through some smart way.

Let us begin by the easy case that is named *affine parametric dependance* where the entries $a(\zeta_i, \zeta_j; \mu)$ appear to read

$$a(\zeta_i, \zeta_j; \mu) = \sum_p g_p(\mu) a_p(\zeta_n, \zeta_m) \,, \tag{48}$$

where the bilinear forms a_p are parameter independent. Many simple problems where the parameter are local constitutive coefficients or local zooming isotropic or non isotropic factors, enter in this framework. The expensive computation of the $a_{p,n,m} = a_p(\zeta_n, \zeta_m)$ can be done offline, once the reduced basis is constructed; these $a_{p,n,m}$ are stored and, for each new problem, the evaluation of the stiffness matrix is done, online, in $P \times N^2$ operations, and solved in $\mathcal{O}(dim X_N^3)$ operations. These figures are coherent with the rapid evaluation of the reduced basis method.

Under the same affine dependance hypothesis on a, it is easy to explain how the a posteriori analysis can be implemented, resulting in a fast on-line solution procedure, provided some off-line computations are made. First of all the computation of $\tilde{\psi}_N$ can be implemented in the space $\tilde{X}_N = \text{Span}\{\xi_1, \ldots, \xi_N)\}$ exactly as above for the computation of u_N. Taking into account (48), together with the expressions obtained from the inversion of problem (38) and (42): $u_N = \sum_{i=1}^N \alpha_i \zeta_i$ and $\tilde{\psi}_N = \sum_{i=1}^N \tilde{\alpha}_i \xi_i$, we can write

$$\mathcal{R}^{pr}(v, \mu) = \sum_p \sum_i g_p(\mu) \alpha_i a_p(\zeta_i, v) - (f, v) \,,$$

and

$$\mathcal{R}^{du}(v, \mu) = -\sum_p \sum_j g_p(\mu) \tilde{\alpha}_j a_p(v, \xi_j) - \ell(v) \,,$$

hence by solving *n*umerically, off-line, each of the problems

$$2\alpha \int \nabla e^{pr,p,i} \nabla v = a_p(\zeta_i, v) \tag{49}$$

$$2\alpha \int \nabla e^{pr,0} \nabla v = (f, v) \tag{50}$$

$$2\alpha \int \nabla e^{du,p,j} \nabla v = a_p(v, \xi_j) \quad (51)$$

$$2\alpha \int \nabla e^{du,0} \nabla v = \ell(v), \quad (52)$$

allows to write the numerical solutions of (46) as a linear combination of the elements previously computed (e.g. $\hat{e}^{pr} = \sum_p \sum_i g_p(\mu)\alpha_i e^{pr,p,i} - e^{pr,0}$) in $\mathcal{O}(PN)$ operations.

References

1. M. Ainsworth, J.T. Oden, A Posteriori Error Estimation in Finite Element Analysis, vol. 37 (Wiley, New York, 2011)
2. I. Babuvska, W.C. Rheinboldt, Error estimates for adaptive finite element computations. SIAM J. Numer. Anal. **15**(4), 736–754 (1978)
3. R. Becker, C. Johnson, R. Rannacher, Adaptive error control for multigrid finite element methods. Computing **55**(4), 271–288 (1995)
4. C. Bernardi, Y. Maday, F. Rapetti, Discrétisations variationnelles de problèmes aux limites elliptiques, vol. 45 (Springer Science and Business Media, 2004)
5. C. Bernardi, B. Métivet, R. Verfüth, Analyse numérique d'indicateurs d'erreur in Maillage et adaptation, ed. by P.-L. George (Hermès, 2001), pp. 251–278
6. P. Binev, A. Cohen, W. Dahmen, R. DeVore, G. Petrova, P. Wojtaszczyk, Convergence rates for greedy algorithms in reduced basis methods. SIAM J. Math. Anal. **43**(3), 1457–1472 (2011)
7. A. Buffa, Y. Maday, A.T. Patera, C. Prudhomme, G. Turinici, A priori convergence of the greedy algorithm for the parametrized reduced basis method. ESAIM: Math. Model. Numer. Anal. **46**(03), 595–603 (2012)
8. C. Canuto, M.Y. Hussaini, A. Quarteroni, T.A. Zang, Spectral Methods (Springer, New York, 2007)
9. P. Clément, Approximation by finite element functions using local regularization, R.A.I.R.O. Anal. Numér. **9**(R2), 77–84 (1975)
10. A. Cohen, R. Devore, Kolmogorov widths under holomorphic mappings. IMA J. Numer. Anal. (2015) to appear
11. R. DeVore, G. Petrova, P. Wojtaszczyk, Greedy algorithms for reduced bases in Banach spaces. Constr. Approx. **37**(3), 455–466 (2013)
12. V. Girault, P.A. Raviart, *Finite Element Approximation of the Navier-Stokes Equations*, Lecture Notes in Mathematics, vol. 749 (Springer, Berlin, 1979)
13. M.A. Grepl, A.T. Patera, A posteriori error bounds for reduced-basis approximations of parametrized parabolic partial differential equations. ESAIM: Math. Model. Numer. Anal. **39**(01), 157–181 (2005)
14. P. Jiránek, Z. Strakos, M. Vohralík, A posteriori error estimates including algebraic error and stopping criteria for iterative solvers. SIAM J. Sci. Comput. **32**(3), 1567–1590 (2010)
15. A. Kolmogoroff, Über die beste Annäherung von Funktionen einer gegebenen Funktionenklasse. Anals Math. **37**, 107–110 (1963)
16. A.T. Patera, E.M. Rønquist, A general output bound result: application to discretization and iteration error estimation and control. Math. Models Methods Appl. Sci. **11**(4), 685–712 (2001)
17. A.T. Patera, G. Rozza, Reduced basis approximation and a posteriori error estimation for parametrized partial differential equations (2007)
18. A. Pinkus, *n-Widths in Approximation Theory* (Springer, Berlin, 1985)
19. C. Prudhomme, D. Rovas, K. Veroy, L. Machiels, Y. Maday, A.T. Patera, G. Turinici, Reliable real-time solution of parametrized partial differential equations: reduced-basis output bound methods. J. Fluids Eng. **124**(1), 70–80 (2002)

20. G. Rozza, D.P. Huynh, A.T. Patera, Reduced basis approximation and a posteriori error estimation for affinely parametrized elliptic coercive partial differential equations. Arch. Comput. Methods Eng. **15**(3), 229–275 (2008)
21. S. Sen, K. Veroy, D.B.P. Huynh, S. Deparis, N.C. Nguyen, A.T. Patera, Natural norm a posteriori error estimators for reduced basis approximations. J. Comput. Phys. (2006)
22. R. Verfürth, A review of a posteriori error estimation techniques for elasticity problems. Comput. Methods Appl. Mech. Eng. **176**(1), 419–440 (1999)
23. R. Verfürth, A posteriori error estimators for convection-diffusion equations. Numerische Mathematik **80**(4), 641–663 (1998)

Implicit Residual Type Error Estimators

Antonio Huerta and Pedro Díez

Abstract The error associated with a numerical solution is intimately related with the residual, that is the lack of verification of the equation by the approximated solution. The residual is computable but obtaining the exact error from the residual is as difficult as computing the exact solution. Residual type estimators provide error assessment tools based on post processing the residual. This post process is either explicit (integrating the residual) or implicit (solving local problems with the residual as source term). Some of the residual type estimates are guaranteed error bounds. The standard estimators aim at assessing the energy norm of the error. Goal-oriented assessment is carried out by considering an auxiliary problem associated with the selected quantity of interest (the adjoint or dual problem). Thus, an error representation allows estimating the error in the quantity of interest as a post-process of the energy measures of the errors in both the original problem and the adjoint one.

Keywords Implicit error estimates · Hybrid-flux equilibration · Flux-free techniques · Goal-oriented estimates · Adjoint problem · Error representation

1 Introduction and Problem Statement

1.1 Preliminaries

The error is the difference between the approximated and the exact solutions, $\boldsymbol{e} := \boldsymbol{u} - \boldsymbol{u}^h$. In the a posteriori error estimation setup, we assume that the approximated solution is available but the exact solution is not. The error is an unknown function

A. Huerta (✉) · P. Díez
Laboratori de Càlcul Numèric, Departament de Matemàtica Aplicada III,
Universitat Politècnica de Catalunya, C/Jordi Girona 1-3, Campus Nord,
Edifici C2, 08034 Barcelona, Spain
e-mail: antonio.huerta@upc.edu

P. Díez
e-mail: pedro.diez@upc.edu

L. Chamoin and P. Díez (eds.), *Verifying Calculations – Forty Years On*,
SpringerBriefs in Applied Sciences and Technology,
DOI 10.1007/978-3-319-20553-3_2

and it is as difficult to obtain as the exact solution. Error estimates aim at providing approximations of (some measures of) the error circumventing the need of having an accurate description of the error function itself.

The equation characterising the error is similar to the original problem, just replacing the source term (loads, in a mechanical context) by the residual associated with the approximated solution. The residual is therefore the driving force for the error. All the information about the error is contained in the residual. However, the key to access this information is the capacity to solve the error equation, which, as previously said, has the same level of complexity and difficulty as the original problem.

The first attempts in a posteriori error assessment aimed at providing approximations of the error measured in energy norm. Lately, a huge effort has been produced in assessing the error in arbitrary quantities of interest. This is of outmost practical importance because it relates with goal-oriented error adaptivity. That is, finding the optimal mesh producing the result specified by the user with the prescribed accuracy at a minimum cost. Moreover, a recently open line of research concentrates in providing certificates of the approximate solution or, conversely, guaranteed upper and lower bounds of the quantity of interest associated with the exact solution.

This article is intended to provide a summary of the different approaches to introduce residual-type error estimates and to highlight the main characteristics of each of them.

1.2 Problem Statement

The standard linear elasticity boundary value problem is used as model problem. The fact of having vectorial unknowns (by opposition to scalar elliptic problems, like the thermal problems modelled by the Poisson equation, which is conceptually similar because has the same mathematical structure) is carrying some additional complexity in the notation but is also allowing us to present points of particular interest that do not appear in the simplest version.

The body under study occupies the domain Ω with boundary $\partial\Omega$, see Fig. 1. The boundary $\partial\Omega$ is divided in two disjoint parts, Γ_N and Γ_D. In the Dirichlet part of the boundary, Γ_D, the displacement is set to be equal to a given value $\boldsymbol{u}_D$. A body load $\boldsymbol{b}$ is applied in Ω and a traction $\boldsymbol{t}$ is applied on the Neumann part of the boundary, Γ_N. The unknown displacement field $\boldsymbol{u}$ and the corresponding stresses $\boldsymbol{\sigma}(\boldsymbol{u})$ are found by solving the following boundary value problem:

$$-\nabla \cdot \boldsymbol{\sigma}(\boldsymbol{u}) = \boldsymbol{b} \qquad \text{in } \Omega, \tag{1a}$$

$$\boldsymbol{\sigma}(\boldsymbol{u}) \cdot \boldsymbol{n} = \boldsymbol{t} \qquad \text{on } \Gamma_N, \tag{1b}$$

$$\boldsymbol{u} = \boldsymbol{u}_D \qquad \text{on } \Gamma_D. \tag{1c}$$

The variational or weak form of problem (1) requires introducing the following functional spaces. The space of admissible displacements $\mathscr{U}$ (a subspace of $\mathscr{H}^1(\Omega)$

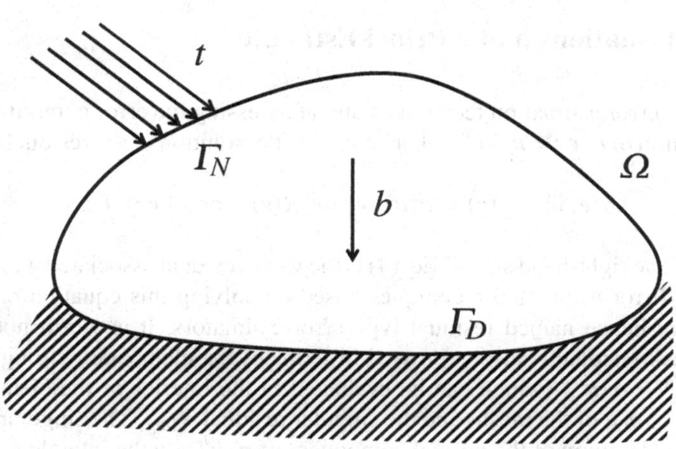

Fig. 1 Illustration of the boundary value problem

of functions fulfilling (1c)) and the space of virtual displacements, $\mathscr{V}$ (also known as trial functions, similar to $\mathscr{U}$ but vanishing on Γ_D). Thus the weak form is readily expressed as find $\boldsymbol{u} \in \mathscr{U}$ such that

$$a(\boldsymbol{u}, \boldsymbol{v}) = l(\boldsymbol{v}), \text{ for all } \boldsymbol{v} \in \mathscr{V}, \tag{2}$$

where

$$a(\boldsymbol{u}, \boldsymbol{v}) := \int_\Omega \boldsymbol{\sigma}(\boldsymbol{u}) : \boldsymbol{\epsilon}(\boldsymbol{v})\, d\Omega \quad , \quad l(\boldsymbol{v}) := \int_\Omega \boldsymbol{b} \cdot \boldsymbol{v}\, d\Omega + \int_{\Gamma_N} \boldsymbol{t} \cdot \boldsymbol{v}\, d\Gamma,$$

being $\boldsymbol{\epsilon}(\cdot)$ the strain operator. Recall that the Hooke tensor $\mathbb{C}$ relates stresses and strains,

$$\boldsymbol{\sigma}(\boldsymbol{u}) = \mathbb{C} : \boldsymbol{\epsilon}(\boldsymbol{u}). \tag{3}$$

It is useful expressing the bilinear form $a(\cdot, \cdot)$ in terms of stresses instead of displacements by formally introducing $\bar{a}(\cdot, \cdot)$ such that

$$\bar{a}(\boldsymbol{\sigma}, \boldsymbol{\tau}) := \int_\Omega \boldsymbol{\sigma} : \mathbb{C}^{-1} : \boldsymbol{\tau}\, d\Omega.$$

Note that, with this definition, $a(\boldsymbol{u}, \boldsymbol{v}) = \bar{a}(\boldsymbol{\sigma}(\boldsymbol{u}), \boldsymbol{\sigma}(\boldsymbol{v}))$.

In this context, the numerical approximation which error is to be assessed, $\boldsymbol{u}^h$ is readily introduced.

A finite element mesh of characteristic size h discretizing Ω induces the functional spaces $\mathscr{U}^h \subset \mathscr{U}$ and $\mathscr{V}^h \subset \mathscr{V}$. The finite element approximation to $\boldsymbol{u}$, $\boldsymbol{u}^h \in \mathscr{U}^h$, is such that

$$a(\boldsymbol{u}^h, \boldsymbol{v}) = l(\boldsymbol{v}), \text{ for all } \boldsymbol{v} \in \mathscr{V}^h.$$

2 Error Equations and a Priori Estimates

A posteriori error estimation techniques aim at assessing the error committed in the approximation of $\boldsymbol{u}$, $\boldsymbol{e} := \boldsymbol{u} - \boldsymbol{u}^h$, where $\boldsymbol{e} \in \mathscr{V}$ is the solution of the residual equation

$$a(\boldsymbol{e}, \boldsymbol{v}) = l(\boldsymbol{v}) - a(\boldsymbol{u}^h, \boldsymbol{v}) =: R(\boldsymbol{v}), \text{ for all } \boldsymbol{v} \in \mathscr{V}. \tag{4}$$

Remark 1 The right-hand side of Eq. (4) is the weak residual associated with the trial function $\boldsymbol{v}$. Error estimation techniques based on solving this equation or making use of it are hence named residual type error estimators. It is worth noting also that the weak residual is also expressed in terms of the elementary strong residual $\boldsymbol{r}_{\text{el}} = \boldsymbol{b} + \nabla \cdot \boldsymbol{\sigma}(\boldsymbol{u}^h)$ (which can be evaluated in the interior of the elements Ω_k, $k = 1, 2, \dots, n_{\text{el}}$, of the mesh) and the singular residual, $\boldsymbol{r}_{\text{sing}}$. The singular residual is defined as the jump of the normal component of $\boldsymbol{\sigma}(\boldsymbol{u}^h)$ on the interelement edges γ (sides in 3D) in Γ_{int}, $\boldsymbol{r}_{\text{sing}} = [\![\boldsymbol{\sigma}(\boldsymbol{u}^h) \cdot \boldsymbol{n}]\!]_{\Gamma_{\text{int}}}$ and as the non verification of the Neumann boundary condition on the element edges γ in Γ_N, $\boldsymbol{r}_{\text{sing}} = \boldsymbol{t} - \boldsymbol{\sigma}(\boldsymbol{u}^h) \cdot \boldsymbol{n}$. The resulting expression is

$$R(\boldsymbol{v}) = \sum_{k=1}^{n_{\text{el}}} \int_{\Omega_k} \boldsymbol{r}_{\text{el}} \cdot \boldsymbol{v} \, d\Omega + \sum_{\gamma \in \Gamma_{\text{int}} \bigcup \Gamma_N} \int_{\gamma} \boldsymbol{r}_{\text{sing}} \cdot \boldsymbol{v} \, d\Gamma. \tag{5}$$

These two components of the residual reveal the existence of two different error sources, the elementary and singular residuals. The former is associated with the lack of verification of the differential equation in the interior of the elements, the latter with the non verification of the continuity requirements of the stress field. The main rationale of the explicit residual error estimates consists in evaluating this two terms separately.

The energy norm of the error, $\|\boldsymbol{e}\|$, is taken as a global measure of the error. This is the norm induced by $a(\cdot, \cdot)$ or $\bar{a}(\cdot, \cdot)$ when applied to stresses, namely

$$\|\boldsymbol{e}\|^2 = a(\boldsymbol{e}, \boldsymbol{e}) = \bar{a}(\boldsymbol{\sigma}_e, \boldsymbol{\sigma}_e) = |\|\boldsymbol{\sigma}_e\||^2,$$

where $\boldsymbol{\sigma}_e$ is the error in stresses $\boldsymbol{\sigma}_e := \boldsymbol{\sigma}(\boldsymbol{u}) - \boldsymbol{\sigma}(\boldsymbol{u}^h)$.

A priori error estimates are expressions of bounds stating the asymptotic behaviour of the error depending of the mesh parameters.

At this point it is interesting reviewing a classical result on interpolation. The values of some function f are known at $n + 1$ sample points x_i, for $i = 0, 1, \dots, n$. The interpolant p is the polynomial of degree n such that $p(x_i) = f(x_i)$, for $i = 0, 1, \dots, n$. The expression of the interpolation error is obtained from the Lagrange interpolation methodology and reads

$$E(x) := f(x) - p(x) = \frac{f^{n+1)}(\nu)}{(n+1)!} L(x)$$

being ν some point in the interval $[x_0, x_n]$ and $L(x)$ a $n+1$ degree polynomial vanishing at the sample points, namely

$$L(x) = \prod_{i=0}^{n}(x - x_i)$$

For a uniform distribution of same points, $h = x_{i+1} - x_i$ for $i = 0, 1, \ldots, n-1$, the particularisation of the previous expression yields

$$E = \frac{f^{n+1)}(\nu)}{4(n+1)} h^{n+1}$$

This expression is used to assess the error associated with the interpolation in the finite element mesh (assuming the exact nodal values are known). In this context, h stands for the characteristic element size of the mesh and n is replaced by p, that is the usual notation for the degree of the polynomial in the finite element approximation. Thus, a finite element interpolation estimate reads

$$\|\boldsymbol{u} - \Pi^h \boldsymbol{u}\|_{\mathscr{L}^2} \leq \underbrace{C \|\boldsymbol{u}\|_{p+1}}_{C^\star} h^{p+1} \tag{6}$$

where Π^h stands for the interpolation operator in the discrete finite element space $\mathscr{V}^h$, $\|\cdot\|_{\mathscr{L}^2}$ is the standard $\mathscr{L}^2$ norm and the norm $\|\cdot\|_{p+1}$ includes the derivatives of order $p+1$ of the argument function. The unknown constant C depends on the geometry of the domain, the regularity of the loading terms and the distortion of the elements in the mesh. It is worth noting that nor C nor the norm $\|\boldsymbol{u}\|_{p+1}$ depend on the mesh parameters h and p. Consequently, the dependence of the (interpolation) error on the mesh parameters is concentrated in the term h^{p+1}. This is the reason of introducing the constant $C^\star$, independent of the mesh parameters.

Starting from Eq. (6) and using the Céa's lemma and the Galerkin orthogonality, the different a priori estimates for the error on the finite element solution are readily obtained. The main idea is replacing $\Pi^h \boldsymbol{u}$ by $\boldsymbol{u}^h$ and generalising the result for norms different than the energy norm.

These expressions are similar to (6) but with different unknown constants and exponential expressions for the dependence on the mesh parameters. Two examples corresponding to the energy (or $\mathscr{H}^1$) norm and the $\mathscr{L}^2$ norm read

$$\|\boldsymbol{u} - \boldsymbol{u}^h\|_{\mathscr{L}^2} \leq \underbrace{C \|\boldsymbol{u}\|_{\mathscr{H}^{p+1}}}_{C^\star} h^{p+1} \tag{7}$$

and

$$\|\boldsymbol{u} - \boldsymbol{u}^h\|_{\mathscr{H}^1} \leq \underbrace{C \|\boldsymbol{u}\|_{\mathscr{H}^{p+1}}}_{C^\star} h^{p} \tag{8}$$

The a priori estimates (7) and (8) are not providing the actual measure of the error but only an expression stating the asymptotic behaviour of the error when the mesh is successively enriched, either increasing p or decreasing h in a uniform manner. For instance, for $p = 1$ (linear elements), one should expect a reduction of the error along a uniform h-refinement which is quadratic if the error is measured in the $\mathscr{L}^2$ norm and only linear if the error is measured in the $\mathscr{H}^1$ norm.

Explicit residual estimates, which are discussed in detail in another chapter, are based on the decomposition of the weak residual discussed in Remark 1. The computable elementary residual $\boldsymbol{r}_{\text{el}}$ and singular residual $\boldsymbol{r}_{\text{sing}}$ are seen as the two sources of error. Explicit estimates are based on postprocessing these two quantities and getting an approximation to the error. Thus, the estimate does not require solving any local problem and is computed directly from the finite element approximation. The input data of the problem to be solved is required to compute the elementary residual, and the part of the singular residual associated with the Neumann boundary. Note that this information is not used in the recovery estimates, which are computed using only $\boldsymbol{\sigma}(\boldsymbol{u}^h)$ and no use is made of the data of the original problem.

The idea of explicit residual estimates uses (5) for $v = e - \Pi^h e$ (Π^h stands for the interpolation operator in $\mathscr{V}^h$) together with the Cauchy-Schwarz inequality and the a priori interpolation estimates. Cooking all these ingredients, the following expression is found (see [1] for a detailed derivation)

$$\|\boldsymbol{e}\|^2 \leq C \left(\sum_{k=1}^{n_{\text{el}}} h_k^2 ||\boldsymbol{r}_{\text{el}}||^2_{\mathscr{L}^2(\Omega_k)} + \sum_{\gamma \in \Gamma_{\text{int}} \bigcup \Gamma_N} h_\gamma ||\boldsymbol{r}_{\text{sing}}||^2_{\mathscr{L}^2(\gamma)} \right), \tag{9}$$

where C is a constant related with the interpolation estimates, generally unknown. Note that each residual is scaled with the local mesh sizes, h_k (element size) and h_γ (edge size), with different exponents arising also from the interpolation estimates. The right-hand side term in (9) is naturally decomposed into elementary contributions and, except for the unknown constant C, it is computable once $\boldsymbol{u}^h$ is obtained.

These estimates are computationally costless and very useful for adaptive procedures where it is important to identify the parts of the domain contributing to the error. Constant C is seen as a single (unknown) multiplicative factor and the local contributions of the elements are therefore properly assessed in a relative basis. Nevertheless, the global value of the error norm is only assessed up to the determination of C. Of course, C could also be estimated and even bounded but in general explicit estimates cannot produce guaranteed upper bounds for $\|\boldsymbol{e}\|$.

3 Implicit Residual Error Estimates

3.1 Classification

Implicit estimators aim at avoiding the disadvantages of explicit estimates by solving the original error equation (4) in a local basis. That is, typically in small domains (the elements or patches of elements) in which a local version of (4) is solved numerically. This requires locally increasing the resolution with respect to the original approximation in $\mathcal{U}^h$. The implicit estimates are classified in different categories, depending on

- the domain in which the local problem is stated: element residual methods (solved element by element) and subdomain residual methods (solved in patches of elements, either centered in nodes or elements)
- the boundary conditions imposed on the local problems: either Dirichlet or Neumann. Roughly speaking, the Dirichlet methods provide continuous approximations to the displacement error and lower bounds of the energy and the Neumann methods yield statically admissible stress fields and upper bounds of the energy error
- the numerical method used to approximate the solution of the local problem: either a standard FE method providing a displacement based approximation of the error (producing the so-called asymptotic estimates which have bounding properties only with respect to a reference solution, not with respect to the exact error) or a dual approach yielding an approximation of the stress field exactly fulfilling the equilibrium equations (producing guaranteed or strict error bounds).

It is worth mentioning here the pioneering work of Ladevèze introducing the error estimators based in the concept of constitutive relation error, see [9]. This family of error estimators is classified here in the implicit residual framework, together with the estimators solving elementary problems with Neumann boundary conditions, because it perfectly matches the category. The rationale for the presentation and the derivation of these techniques is however pretty different. Following this line of thought, based also in mechanical arguments, strategies to generalize these tools to nonlinear and transient problems have been suggested, see [5].

An alternative approach fitting also the implicit residual philosophy are the so-called *dual global solvers*. This strategy is based on the ideas introduced by [8]. A statically admissible stress field $\boldsymbol{\sigma}^\star$ is obtained by means of a global computation over a discrete space $\mathcal{S}^h$ (where the stresses are interpolated). This requires solving a global optimization problem reading: find $\boldsymbol{\sigma}^\star \in \mathcal{S}^h$ such that the complementary energy $|||\boldsymbol{\sigma}^\star|||^2 = \bar{a}(\boldsymbol{\sigma}^\star, \boldsymbol{\sigma}^\star)$ is minimum, with the additional restriction of being statically admissible, see (10). Thus, the statically admissible stress field $\boldsymbol{\sigma}^\star$ produces an upper bound energy norm estimate, overestimating $\|\boldsymbol{e}\|$, see Sect. 3.2. Moreover, this error bound is the sharper you can get in $\mathcal{S}^h$. Thus, estimates based on *dual global solvers* are generally sharp. Nevertheless, the global nature of the dual approximation makes them computationally expensive. Both the element residual methods and the subdomain residual methods are alternatives based on solving only local problems and, consequently, providing upper bounds of the error at an affordable computational cost.

3.2 Assessing the Energy Norm of the Error

A first step in a posteriori assessment is estimating the error measured in the energy norm, that is obtaining a good approximation of $\boldsymbol{\sigma}_e$ and computing $\|\boldsymbol{e}\|$. This translates in finding a new stress field $\boldsymbol{\sigma}^\star$ based on the information at hand, that is $\boldsymbol{\sigma}(\boldsymbol{u}^h)$, and such that $\boldsymbol{\sigma}^\star$ approximates the actual stresses $\boldsymbol{\sigma}(\boldsymbol{u})$ much better than $\boldsymbol{\sigma}(\boldsymbol{u}^h)$. Thus, a computable error estimate is readily obtained

$$\boldsymbol{\sigma}_e \approx \boldsymbol{\sigma}_e^\star = \boldsymbol{\sigma}^\star - \boldsymbol{\sigma}(\boldsymbol{u}^h),$$

yielding also the corresponding energy norm estimate $|||\boldsymbol{\sigma}_e^\star|||^2 = \bar{a}(\boldsymbol{\sigma}_e^\star, \boldsymbol{\sigma}_e^\star) \approx \|\boldsymbol{e}\|^2$.

The stress field $\boldsymbol{\sigma}^\star$ is said to be statically admissible if it is continuous (at least in the normal components to the discontinuity surface, that is without traction jumps) and it fulfills the equilibrium equations (1a) and (1b). This is equivalent to say that for all the virtual displacements $\boldsymbol{v} \in \mathscr{V}$

$$\bar{a}\big(\boldsymbol{\sigma}^\star, \boldsymbol{\sigma}(\boldsymbol{v})\big) = l\big(\boldsymbol{v}\big). \tag{10}$$

Note that the solution of (10) is not unique because $\boldsymbol{\sigma}^\star$ is not assumed to fulfill any compatibility condition, in other words $\boldsymbol{\sigma}^\star$ does not necessarily derive from a displacement field following (3).

A statically admissible stress field $\boldsymbol{\sigma}^\star$ produces an energy norm estimate $|||\boldsymbol{\sigma}_e^\star|||$ larger than (or equal to) $\|\boldsymbol{e}\|$. The error estimation technique providing this kind of error approximation is referred as an upper bound error estimator. The upper bound property of the statically admissible stress field is readily derived by considering $\boldsymbol{v} = \boldsymbol{e}$ in (2) and (10), thus

$$\bar{a}\big(\boldsymbol{\sigma}(\boldsymbol{u}), \boldsymbol{\sigma}_e\big) = l\big(\boldsymbol{e}\big) = \bar{a}\big(\boldsymbol{\sigma}^\star, \boldsymbol{\sigma}_e\big)$$

and subtracting $\bar{a}\big(\boldsymbol{\sigma}(\boldsymbol{u}^h), \boldsymbol{\sigma}_e\big)$ in both sides

$$\bar{a}\big(\boldsymbol{\sigma}_e, \boldsymbol{\sigma}_e\big) = \bar{a}\big(\boldsymbol{\sigma}_e^\star, \boldsymbol{\sigma}_e\big),$$

which yields $|||\boldsymbol{\sigma}_e||| \leq |||\boldsymbol{\sigma}_e^\star|||$ by simply considering the Cauchy-Schwarz inequality.

Thus, the key issue in any error estimation technique is to produce a properly enhanced stress field $\boldsymbol{\sigma}^\star$. Moreover, if $\boldsymbol{\sigma}^\star$ is build up such that it is statically admissible, then this additional feature confers to the estimator the upper bound property. The strategies producing the enhanced stresses $\boldsymbol{\sigma}^\star$ are classified into two categories: recovery type estimators (no discussed here) and implicit residual type estimators.

It is worth remarking that, in general, the enhanced stress $\boldsymbol{\sigma}^\star$ and the corresponding stress error $\boldsymbol{\sigma}_e$ can only be used to evaluate the energy norm of the error, and no other quantities. In particular, any magnitude based on the displacement error cannot be directly evaluated using $\boldsymbol{\sigma}^\star$.

3.3 Element Residual Method; Equilibrated Residual Estimates

The local version of the error equation (4) in the element Ω_k of the mesh states that the restriction of the error $\boldsymbol{e}$ to Ω_k fulfills

$$a_k(\boldsymbol{e},\boldsymbol{v}) = l_k(\boldsymbol{v}) - a_k(\boldsymbol{u}^h,\boldsymbol{v}) + \int_{\partial\Omega_k\setminus\partial\Omega} (\boldsymbol{\sigma}(\boldsymbol{u})\cdot\boldsymbol{n})\cdot\boldsymbol{v}\,d\Gamma \tag{11}$$

for all v taking values in Ω_k. Subscript k in the linear and bilinear forms indicates that the corresponding integrals are restricted to Ω_k. Note that the last term of the right-hand side stands for the local Neumann boundary conditions and depends on the unknown traction associated with the exact solution. Note also that the local error stress field $\boldsymbol{\sigma}(\boldsymbol{e})$ fulfills a variant of (11), substituting the left-hand side term by $\bar{a}_k\big(\boldsymbol{\sigma}(\boldsymbol{e}),\boldsymbol{\sigma}(\boldsymbol{v})\big)$.

In order to obtain a solvable local problem, the unknown boundary traction $\boldsymbol{\sigma}(\boldsymbol{u})\cdot\boldsymbol{n}$ on the boundary of Ω_k is replaced by some approximated value $\boldsymbol{g}_k$ that has to be determined on all the interelement edges. Thus, the local equation for the approximated stress error, $\boldsymbol{\sigma}_e^\star$ is

$$\bar{a}_k\big(\boldsymbol{\sigma}_e^\star,\boldsymbol{\sigma}(\boldsymbol{v})\big) = l_k(\boldsymbol{v}) - a_k(\boldsymbol{u}^h,\boldsymbol{v}) + \int_{\partial\Omega_k\setminus\partial\Omega} \boldsymbol{g}_k\cdot\boldsymbol{v}\,d\Gamma. \tag{12}$$

In order to provide statically admissible stresses, the approximated traction $\boldsymbol{g}_k$ has to fulfill two properties

1. on the common edge of two contiguous elements Ω_k and $\Omega_{k'}$, $\boldsymbol{g}_k = -\boldsymbol{g}_{k'}$ (this is to guarantee the continuity of the traction associated with $\boldsymbol{\sigma}^\star$)
2. the boundary traction must be in equilibrium with the interior loads. This *compatibility condition* is needed to ensure that the problem (12) is solvable.

The compatibility condition requires $\boldsymbol{g}_k$ to fulfill

$$l_k(\boldsymbol{v}) - a_k(\boldsymbol{u}^h,\boldsymbol{v}) + \int_{\partial\Omega_k\setminus\partial\Omega} \boldsymbol{g}_k\cdot\boldsymbol{v}\,d\Gamma = 0 \tag{13}$$

for any rigid body motion $\boldsymbol{v}$ (in 2D, this means $\boldsymbol{v}$ taking the values of the two translations $\boldsymbol{t}_x$ and $\boldsymbol{t}_y$ and the rotation $\boldsymbol{\theta}$). If this condition is fulfilled, problem (12) is solvable (the solution exists, even if it is not unique). Any of the solutions of this problem produces an upper bound estimate.

The first idea to determine $\boldsymbol{g}_k$ was introduced by [3] and consists in taking $\boldsymbol{g}_k$ equal to the average of the numerical normal traction, computed from $\boldsymbol{\sigma}(\boldsymbol{u}^h)$. This is equivalent to assume that $\boldsymbol{\sigma}(\boldsymbol{u})\cdot\boldsymbol{n} \approx \langle\boldsymbol{\sigma}(\boldsymbol{u}^h)\rangle_{\text{ave}}\cdot\boldsymbol{n}$ on the interelement edges. This option fulfills the continuity restriction but fails guaranteeing the compatibility condition (13). To overcome this problem, [3] propose the following work-around: the test function $\boldsymbol{v}$ in problem (12) is taken in a restricted functional space of functions

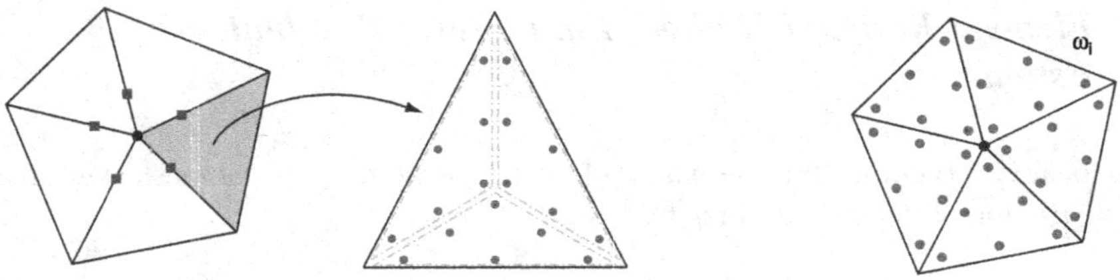

Fig. 2 Illustration of the element residual method (*left*) and the subdomain residual method (*right*). In the element residual method the contribution to the tractions $\boldsymbol{g}_k$ of every node of the mesh (represented by the *blue squares*) are computed in a nodal basis. Then, the tractions $\boldsymbol{g}_k$ are used to solve the local elementary problems and the stresses inside the element fulfilling the equilibrium are determined. Subdomain residual method: A larger local problem is solved for each node of the mesh but no equilibrated tractions have to be computed. The *red circles* represent the degrees of freedom describing the approximated stresses. (color in online)

vanishing at the vertex nodes of element Ω_k. This simple approach only yields statically admissible estimates stresses $\boldsymbol{\sigma}_e^\star$ if the error on the nodes of the mesh is zero. This is not the general case and consequently if this strategy is used, the upper bound property cannot be guaranteed. An alternative also devised in [3] consists in replacing in the right-hand side of (12) $\boldsymbol{v}$ by $\boldsymbol{v} - \Pi^h\boldsymbol{v}$. This automatically guarantees that the local problem is compatible (or equilibrated) and preserves the global upper bound property. The global property is kept because subtracting $\Pi^h\boldsymbol{v}$ in the argument of $R(\cdot)$ (the right-hand side of (4)) does not change the error equation. This smart operation can also be seen as an implicit way of recovering a compatible traction $\boldsymbol{g}_k$.

This is the basis of the so-called *equilibrated residual estimates*. In fact, this family of estimators introduces efficient and practical algorithms for constructing equilibrated fluxes, that is recovering $\boldsymbol{g}_k$ by solving only local problems. The compatibility condition (13) is at the first sight a global restriction, involving the tractions on all the element boundaries. If the equilibrated residual methods are among the most popular implicit residual type estimators is because the computation of the tractions $\boldsymbol{g}_k$ is decoupled node to node. Using a smart representation of $\boldsymbol{g}_k$, the nodal contributions to $\boldsymbol{g}_k$ on all the edges converging in a given node are computed independently, and it requires solving a small linear system of equations as indicated in [2, 9–11, 14, 16] (Fig. 2).

3.4 Subdomain Residual Methods; Flux-Free Estimates

The effectivity of the equilibrated residual method depends on the quality of the local tractions $\boldsymbol{g}_k$. For instance, the dual-global estimates are usually much sharper than the equilibrated residual estimates. Moreover, although computing $\boldsymbol{g}_k$ as indicated above is computationally inexpensive because the local problems are decoupled, the implementation of the equilibration techniques is often involved and difficult to generalize to different element types or space dimensions.

The subdomain residual methods are introduced as an alternative to equilibrated residual methods such that:

- they preclude solving a global problem (the local equations are posed in different subdomains, patches of elements surrounding a node, also denoted as *stars*)
- they provide upper bound estimates
- they circumvent the necessity of finding proper tractions as boundary conditions for the local problems. The local boundary conditions are *natural* and the estimates are also said to be *flux-free*.

In order to localize the error equation (4), use is made of the partition of unity property. Let ϕ_i be the linear finite element interpolation function associated with the ith vertex node of the mesh. Note that these functions sum up to the unity and that the support of ϕ_i is precisely the patch of elements containing this node, ω_i. Thus, a local version of (4) in ω_i, providing a local approximation $\boldsymbol{\sigma}_e^{\star i}$ of the stress error, is readily recovered as

$$\bar{a}_{\omega^i}\left(\boldsymbol{\sigma}_e^{\star i}, \boldsymbol{\sigma}(\boldsymbol{v})\right) = R(\phi_i \boldsymbol{v}) \tag{14}$$

for all $\boldsymbol{v}$ taking values in ω_i, being $\bar{a}_{\omega^i}(\cdot,\cdot)$ the restriction of $\bar{a}(\cdot,\cdot)$ to ω_i. The sum of the local approximations to the stress error $\boldsymbol{\sigma}_e^{\star i}$ provide a statically admissible stress field $\boldsymbol{\sigma}_e$ and its corresponding error norm is a sharp upper bound of the error, see [15]. The local problem (14) is automatically equilibrated in most of the cases because the right-hand side vanishes for $\boldsymbol{v}$ equal to a rigid body motion. In the unique case in which this equilibrium is not automatically guaranteed (linear elements for structural mechanics) a straightforward modification is introduced to ensure solvability, see [15].

Similar approaches are developed taking $\bar{a}_{\omega^i}(\cdot,\cdot)$ as a locally weighted version of $\bar{a}(\cdot,\cdot)$, see [4, 12, 13]. In this case the upper bound estimate is obtained adding the squared norms of the local contributions rather than adding the functions and computing the norm afterwards. The estimates obtained following this rationale are not as sharp as the ones obtained taking $\bar{a}_{\omega^i}(\cdot,\cdot)$ as simple restriction of $\bar{a}(\cdot,\cdot)$.

4 Goal-Oriented Estimates

Assessing the energy norm of the error is not sufficient for many applications. In practice, the finite element user is interested in specific magnitudes extracted from the global solution by some post-process. These magnitudes are referred as *quantities of interest* or *functional outputs*. Goal-oriented error assessment strategies aim at estimating the error committed in these quantities and possibly providing bounds for it.

The quantities of interest considered here are linear functional outputs of the solution, $l^{\mathcal{O}}(\boldsymbol{u})$. In particular, those expressed in the form

$$l^{\mathcal{O}}(\boldsymbol{u}) = \int_{\Omega} \boldsymbol{b}^{\mathcal{O}} \cdot \boldsymbol{u} \, d\Omega + \int_{\Gamma_N} \boldsymbol{t}^{\mathcal{O}} \cdot \boldsymbol{u} \, d\Gamma + a(\boldsymbol{u}, \boldsymbol{\chi}^{\mathcal{O}}), \tag{15}$$

where $\boldsymbol{b}^{\mathcal{O}}$, $\boldsymbol{t}^{\mathcal{O}}$ and $\boldsymbol{\chi}^{\mathcal{O}}$ are given functions characterizing the quantity of interest. Note that $l^{\mathcal{O}}(\cdot)$ has the same structure as the right-hand side of (2). The extension to nonlinear outputs is discussed in [17].

This expression is pretty general and accounts for a large variety of quantities of interest. The first term in (15) is a weighted average of the displacements, being $\boldsymbol{b}^{\mathcal{O}}$ the weight. Note that this average is restricted to the support of $\boldsymbol{b}^{\mathcal{O}}$ which is in practice the way of indicating the zone of interest. Similarly, the second term in (15) accounts for averaged displacements along a part of the Neumann boundary. Note that displacements on the Dirichlet boundary, Γ_D, are known a priori and therefore it makes not sense to include in the quantity of interest averaged displacements on Γ_D. On the contrary, tractions on Γ_D are generally interesting for the end-users, as they are reaction forces on the supports. In fact, this kind of quantities are accounted by the third term in (15). At first sight, the third term in (15) only represents an average of the stresses in the interior of the domain of study. However, a proper choice of function $\boldsymbol{\chi}^{\mathcal{O}}$ allows also representing traction averages along Γ_D. This is readily demonstrated by noting that

$$\int_{\Gamma_D} \boldsymbol{\chi}^{\mathcal{O}} \cdot (\boldsymbol{\sigma}(\boldsymbol{u}) \cdot \boldsymbol{n}) \, d\Gamma = a(\boldsymbol{u}, \boldsymbol{\chi}^{\mathcal{O}}) - \int_{\Gamma_N} \boldsymbol{t} \cdot \boldsymbol{\chi}^{\mathcal{O}} \, d\Gamma - \int_{\Omega} \boldsymbol{b} \cdot \boldsymbol{\chi}^{\mathcal{O}} \, d\Omega. \tag{16}$$

Equation (16) is obtained after the usual algebraic manipulation, using the weighted residuals technique into the original equation (1), taking $\boldsymbol{\chi}^{\mathcal{O}}$, which does not vanish on Γ_D, as weighting function. It is clear from (16) that the third term in (15) is a traction average on Γ_D plus a computable term involving part of the data.

The expression (15) allows also determining pointwise quantities by using functions of the Dirac delta type although in practice smeared versions are preferred (averages in neighborhoods of the point) in order to avoid singularities.

The objective of the goal-oriented error assessment is to estimate the value of $l^{\mathcal{O}}(\boldsymbol{e})$ which, for linear outputs, coincides with $l^{\mathcal{O}}(\boldsymbol{u}) - l^{\mathcal{O}}(\boldsymbol{u}_H)$.

As pointed out in the previous section, the enhanced stresses $\boldsymbol{\sigma}^{\star}$ can only be used to assess the energy norm of the error. Thus, an error representation is needed to express the error in the quantity of interest in terms of the energy error. This error representation requires introducing an auxiliary problem, denoted as *adjoint* or *dual* problem by different authors. This problem reads: find $\boldsymbol{\psi} \in \mathcal{V}$ such that

$$a(\boldsymbol{v}, \boldsymbol{\psi}) = l^{\mathcal{O}}(\boldsymbol{v}), \text{ for all} \boldsymbol{v} \in \mathcal{V}. \tag{17}$$

Note that the adjoint solution $\boldsymbol{\psi}$ lies in the space $\mathcal{V}$ (that is vanishes on Γ_D) and that, for the sake of clarity, the order of the arguments in $a(\cdot, \cdot)$ is switched with respect to the original equation (2). The numerical solution of the adjoint problem (17), $\boldsymbol{\psi}^h$, has the associated error $\boldsymbol{\varepsilon} := \boldsymbol{\psi} - \boldsymbol{\psi}^h$. These auxiliary functions are introduced such that the following error representation holds:

$$l^{\mathcal{O}}(e) = a(e, \psi) = a(e, \varepsilon).$$

This error representation allows bounding the error in terms of the energy norm of the errors in the direct and adjoint problem. This is a direct consequence of the Cauchy-Schwarz inequality, namely

$$|l^{\mathcal{O}}(e)| = |a(e, \varepsilon)| \leq \|e\| \, \|\varepsilon\|. \tag{18}$$

An obvious error bound for the quantity of interest follows: $l^{\mathcal{O}}(e)$ ranges between $\pm\|e\| \, \|\varepsilon\|$. Thus, an upper bound of the quantity of interest (in absolute value) is obtained if upper bounds for $\|e\|$ and $\|\varepsilon\|$ are available. The sharpness of this upper and lower bounding of the error in the quantity of interest is improved by considering the so-called parallelogram identity:

$$l^{\mathcal{O}}(e) = \frac{1}{4}\|\kappa e + \frac{1}{\kappa}\varepsilon\|^2 - \frac{1}{4}\|\kappa e - \frac{1}{\kappa}\varepsilon\|^2 \tag{19}$$

standing for any non-zero factor κ. It follows from (19) that an upper bound for $l^{\mathcal{O}}(e)$ is obtained by combining an upper bound for $\|\kappa e + \frac{1}{\kappa}\varepsilon\|$ and a lower bound for $\|\kappa e - \frac{1}{\kappa}\varepsilon\|$ (using zero as a lower bound is a not sharp but robust option). Conversely a lower bound for $l^{\mathcal{O}}(e)$ is obtained by combining a lower bound for $\|\kappa e + \frac{1}{\kappa}\varepsilon\|$ and an upper bound for $\|\kappa e - \frac{1}{\kappa}\varepsilon\|$. In practice, if the lower bounds are properly assessed, this alternative is much sharper than using only (18) and usually allows determining the sign of $l^{\mathcal{O}}(e)$ because both upper and lower bounds may have the same sign.

4.1 Lower Bounds for the Energy Using Implicit Dirichlet Estimates

Recall that in order to get sharp bounds of the error in the quantities of interest using (19), it is important to obtain lower bounds of the energy norm of the error. Any continuous approximation of the displacement error, $e^\star \in \mathcal{V}$, is such that $R(e^\star)\|e^\star\|^{-1} \leq \|e\|$. This is a direct consequence of taking $v = e^\star$ in (4) (this is only possible if $e^\star$ is continuous) and use the Cauchy-Schwarz inequality. Thus, a lower bound is easily recovered after $e^\star$.

The simplest way of guaranteeing continuity by solving local residual problems is to use homogeneous Dirichlet boundary conditions (prescribe displacements equal to zero) on the boundary of the local subdomains. This idea was used in [6] solving such problems elementwise and then complementing the estimate by adding the contribution of a new family of subdomains overlapping the elements while keeping the lower bound property in the resulting error assessment.

The continuous estimate $\boldsymbol{e}^\star$ can also be obtained using the recovery techniques or postprocessing the local solution of the residual type estimates based on Neumann local problems as described in [7]. Obviously, the quality of the resulting lower bound depends on how well $\boldsymbol{e}^\star$ approximates $\boldsymbol{e}$, in particular, for $\boldsymbol{e}^\star = \boldsymbol{e}$, $R(\boldsymbol{e}^\star)\|\boldsymbol{e}^\star\|^{-1} = \|\boldsymbol{e}\|$ and the estimate is therefore exact.

Note that the energy norm assessment and energy bounds for the direct (or primal) and adjoint problems (or the combined problems yielding $\kappa \boldsymbol{e} \pm \frac{1}{\kappa}\boldsymbol{\varepsilon}$) are the basic underlying tools for goal oriented assessment.

References

1. M. Ainsworth, J.T. Oden, A posteriori error estimation in finite element analysis, in *Pure and Applied Mathematics (New York)* (Wiley-Interscience [Wiley], Chichester, 2000)
2. M. Ainsworth, J.T. Oden, A unified approach to a posteriori error estimation using element residual methods. Numer. Math. **65**(1), 23–50 (1993)
3. R.E. Bank, A. Weiser, Some a posteriori error estimators for elliptic partial differential equations. Math. Comput. **170**(44), 283–301 (1985)
4. C. Carstensen, S.A. Funken, Fully reliable localized error control in the FEM. SIAM J. Sci. Comput. **4**(21), 1465–1484 (1999–2000)
5. L. Chamoin, P. Ladevèze, A non-intrusive method for the calculation of strict and efficient bounds of calculated outputs of interest in linear viscoelasticity problems. Comput. Methods Appl. Mech. Eng. **197**(9–12), 994–1014 (2008)
6. P. Díez, J.J. Egozcue, A. Huerta, A posteriori error estimation for standard finite element analysis. Comput. Methods Appl. Mech. Eng. **163**(1–4), 141–157 (1998)
7. P. Díez, N. Parés, A. Huerta, Recovering lower bounds of the error by postprocessing implicit residual a posteriori error estimates. Int. J. Numer. Methods Eng. **10**(56), 1465–1488 (2003)
8. B. Fraeijs e Veubeke, Displacement and equilibrium models in the finite element method, in *Zienkiewicz and Holister, Editors, Stress Analysis* (Wiley, London, 1965). Int. J. Numer. Methods Eng., Classical Reprint Series **52**, 287–342 (2001)
9. P. Ladevèze, D. Leguillon, Error estimate procedure in the finite element method and applications. SIAM J. Numer. Anal. **20**(3), 485–509 (1983)
10. P. Ladevèze, E.A.W. Maunder, A general method for recovering equilibrating element tractions. Comput. Methods Appl. Mech. Eng. **137**(41), 111–151 (1996)
11. P. Ladevèze, J.-P. Pelle, P. Rougeot, Error estimation and mesh optimization for classical finite elements Engineering Computations. Int. J. Comput.-Aided Eng. Softw. **8**(1), 69–80 (1991)
12. L. Machiels, Y. Maday, A.T. Patera, rite authors, A "flux-free" nodal Neumann subproblem approach to output bounds for partial differential equations. C.R. Acad. Sci. Série I. Math. **3**(330), 249–254 (2000)
13. P. Morin, R.H. Nochetto, K.G. Siebert, Local problems on stars: a posteriori error estimators, convergence, and performance. Math. Comput. **243**(72), 1067–1097 (2003)
14. N. Parés, J. Bonet, A. Huerta, J. Peraire, The computation of bounds for linear-functional outputs of weak solutions to the two-dimensional elasticity equations. Comput. Methods Appl. Mech. Eng. **195**(4–6), 406–429 (2006)
15. N. Parés, P. Díez, A. Huerta, Subdomain-based flux-free a posteriori error estimators. Comput. Methods Appl. Mech. Eng. **195**(4–6), 297–323 (2006)
16. A.M. Sauer-Budge, J. Bonet, A. Huerta, J. Peraire, Computing bounds for linear functionals of exact weak solutions to Poisson's equation. SIAM J. Numer. Anal. **42**(4), 1610–1630 (2004)
17. Z.C. Xuan, N. Parés, J. Peraire, Computing upper and lower bounds for the J-integral in two-dimensional linear elasticity. Comput. Methods Appl. Mech. Eng. **195**(4–6), 430–443 (2006)

Fundaments of Recovery-Based Error Estimation and Bounding

E. Nadal and J.J. Ródenas

Abstract Several error estimators have been developed in order to control the accuracy of the Finite Element simulations. Most of these techniques can be divided in three main groups: (i) residual-based error estimators, (ii) dual techniques and (iii) recovery-based error estimators. In this work we will introduce the main ideas of the recovery-based error estimators.

Keywords Error estimation · Super-convergent Patch Recovery · Recovery-based error estimators · Error bounding · Error in Quantities of Interest

1 Introduction

During last decades several a posteriori error estimation approaches have been developed by researches. Among them we can highlight the residual (explicit or implicit) error estimators, the error estimation techniques based on dual approaches and the recovery-based error estimators. In this work we will introduce the recovery-based error estimators. The measure of the error tries to indicate the difference between the exact solution and the approximated solution obtained by a numerical method such as the Finite Element Method (FEM). This error has to be measured by a norm. The energy norm has been traditionally used, however, norms based in Quantities of Interest (QoI) have also been recently introduced.

Let us first introduce the 2D linear elasticity problem. This problem will be used for all the numerical experiments. We denote the Cauchy stress as $\boldsymbol{\sigma}$, the displacement as $\boldsymbol{u}$ and the strain as $\boldsymbol{\varepsilon}$, all these fields being defined over the domain $\Omega \subset \mathbb{R}^2$, with boundary denoted by $\partial\Omega$. Prescribed tractions denoted by $\boldsymbol{t}$ are imposed over the

E. Nadal (✉)
École Centrale de Nantes (ECN). 1, Rue de la Noë, 44300 Nantes, France
e-mail: Enrique.Nadal@ec-nantes.fr

J.J. Ródenas
Universitat Politècnica de València, Camino de Vera s/n, 46022 Valencia, Spain
e-mail: jjrodena@mcm.upv.es

L. Chamoin and P. Díez (eds.), *Verifying Calculations – Forty Years On*,
SpringerBriefs in Applied Sciences and Technology,
DOI 10.1007/978-3-319-20553-3_3

part Γ_N of the boundary, while displacements denoted by $\bar{\boldsymbol{u}}$ are prescribed over the complementary part Γ_D of the boundary. Body loads are denoted as $\boldsymbol{b}$.

The elasticity problem takes the following form. We seek $(\boldsymbol{\sigma}, \boldsymbol{u})$ satisfying:

$$\boldsymbol{L}^T\boldsymbol{\sigma} + \boldsymbol{b} = \boldsymbol{0} \quad \text{in } \Omega \tag{1}$$

$$\boldsymbol{\sigma} = \boldsymbol{D}\boldsymbol{\varepsilon}(\boldsymbol{u}) \quad \text{in } \Omega \tag{2}$$

$$\boldsymbol{G}\boldsymbol{\sigma} = \boldsymbol{t} \quad \text{on } \Gamma_N \tag{3}$$

$$\boldsymbol{u} = \bar{\boldsymbol{u}} \quad \text{on } \Gamma_D \tag{4}$$

where $\boldsymbol{D}$ is the Hook's tensor relating the stresses with the strains, $\boldsymbol{G}$ is the projection operator that projects the stress field into traction over the boundary and $\boldsymbol{L}$ is the differential operator defined as:

$$\boldsymbol{L} = \begin{bmatrix} \frac{\partial}{\partial x} & 0 \\ 0 & \frac{\partial}{\partial y} \\ \frac{\partial}{\partial y} & \frac{\partial}{\partial x} \end{bmatrix} \tag{5}$$

The problem above takes the following primal variational form:

$$\begin{aligned} &\text{Find } \boldsymbol{u} \in (V + \{\boldsymbol{w}\}) : \forall \boldsymbol{v} \in V \\ &a(\boldsymbol{u}, \boldsymbol{v}) = l(\boldsymbol{v}) \\ &a(\boldsymbol{u}, \boldsymbol{v}) = \int_\Omega \boldsymbol{\varepsilon}(\boldsymbol{u})^T \boldsymbol{D}\boldsymbol{\varepsilon}(\boldsymbol{v})\, \mathrm{d}\Omega \\ &l(\boldsymbol{v}) = \int_\Omega \boldsymbol{b}^T \boldsymbol{v}\, \mathrm{d}\Omega + \int_{\Gamma_N} \boldsymbol{t}^T \boldsymbol{v}\, \mathrm{d}\Gamma \end{aligned} \tag{6}$$

where $V = \{\boldsymbol{v} \mid \boldsymbol{v} \in \left[H^1(\Omega)\right]^2, \boldsymbol{v}|_{\Gamma_D} = \boldsymbol{0}\}$ and $\boldsymbol{w}$ is a particular displacement field satisfying the Dirichlet boundary conditions. The corresponding FE discretization can be written in the following manner. The approximate displacement field $\boldsymbol{u}^h$ is searched for in a space of finite dimension $(V^h + \{\boldsymbol{w}\}) \subset (V + \{\boldsymbol{w}\})$ such that V^h is spanned by locally supported finite element shape functions.

Using the Galerkin framework, the primal variational formulation (1) and (2) is recast in the form:

$$\begin{aligned} &\text{Find } \boldsymbol{u}^h \in \left(V^h + \{\boldsymbol{w}\}\right) : \forall \boldsymbol{v} \in V^h \\ &a(\boldsymbol{u}^h, \boldsymbol{v}) = l(\boldsymbol{v}) \end{aligned} \tag{7}$$

which can be solved using classical finite element technology [1]. Since the FE solution is an approximation to the exact solution, the exact error in the energy norm of the FE solution can be written as:

$$|||\boldsymbol{e}|||^2_\Omega = \int_\Omega (\boldsymbol{\sigma} - \boldsymbol{\sigma}^h)^T \boldsymbol{D}^{-1} (\boldsymbol{\sigma} - \boldsymbol{\sigma}^h)\ \mathrm{d}\Omega \tag{8}$$

where $\boldsymbol{e} = \boldsymbol{u} - \boldsymbol{u}^h$. Zienkiewicz and Zhu [2] introduced the so-called Zienkiewicz and Zhu (ZZ) error estimator (9) which consists of substituting the unknown field $\boldsymbol{\sigma}$ in expression (8) by the recovered stress field $\boldsymbol{\sigma}^*$, more accurate than $\boldsymbol{\sigma}^h$, evaluated with a recovery procedure, leading to:

$$\mathscr{E}^2 := \int_{\Omega} (\boldsymbol{\sigma}^* - \boldsymbol{\sigma}^h)^T \boldsymbol{D}^{-1} (\boldsymbol{\sigma}^* - \boldsymbol{\sigma}^h)\ \mathrm{d}\Omega \tag{9}$$

In its original form the estimator $\mathscr{E}$ did not intended to be an upper bound of the true error but an approximation whose accuracy was directly related to the hability of $\boldsymbol{\sigma}^*$ to accurately represent $\boldsymbol{\sigma}$. There are several recovery procedures to obtain $\boldsymbol{\sigma}^*$ described in literature from which we can highlight the *nodal averaging* technique which is the simplest one. The recovered stress field provided by this technique is obtained by a nodal representation of the stress field. The value assigned to each node is obtained as an average of the stress values evaluated at the node from each of the elements connected to the node. The FE shape functions are then used to interpolate a continuous recovered stress field. This technique is extremely simple and allows for fast calculations, providing acceptable results for linear elements. However, for quadratic elements, it does not provide good error estimations since the effectivity index θ does not converge to the ideal value ($\theta = \frac{\mathscr{E}}{|||\boldsymbol{e}|||_{\Omega}} \not\to 1$). In literature we can find a variety of recovery processes based on the same nodal representation scheme but guaranteeing the *asymptotically exact* property ($\theta \to 1$ with mesh refinement). Among them, we can highlight the Super-convergent Patch Recovery (SPR) technique introduced by Zienkiewicz and Zhu [3, 4].

In the remaining, in Sect. 2, we will introduce advanced recovery techniques for error estimation purposes. Section 3 will be devoted to different techniques to obtain error bounds in energy norm. Section 4 will introduce a novel technique to estimate the error not of the FE solution, but of the recovered solution and finally Sect. 5 will provide some numerical evidences to show the quality of the proposed methods.

2 Recovery-Based Error Estimators

2.1 The SPR Technique

The Super-convergent Patch Recovery (SPR) technique provides a recovered stress field of a better quality providing *asymptotically exact* error estimates for practical situations [5–7]. The SPR technique is widely used to obtain the improved stress field $\boldsymbol{\sigma}^*$ used in the ZZ error estimator (9). References [8–10] show that the SPR technique is the most robust technique used for error estimation on problems with smooth solutions approximated on patch-wise uniform grids, for linear or quadratic elements. The recovery procedure consists first of defining a patch of elements $\mathscr{P}^i$, that is a set of elements sharing a vertex node $i \in \mathscr{N}$, being $\mathscr{N}$ the number of nodes in the mesh. This node is also called the patch assembly node, see Fig. 1a.

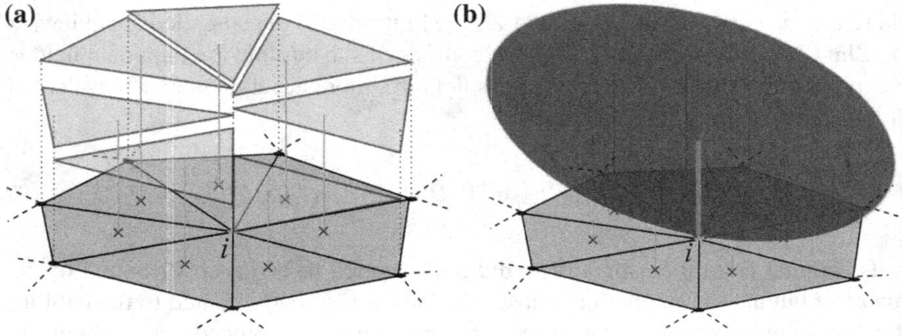

Fig. 1 Recovery in a patch of linear triangular elements. The *black points* are the nodes of the mesh. The *red node* is the patch assembly node. The super-convergent points are indicated by *blue crosses*. **a** FE stress field $\boldsymbol{\sigma}^h$ (transparent surfaces) in the patch. **b** Least squares fitted polynomial surface. The *pink line* represents the stress value at the patch assembly node, the only value that is retained in the standard SPR. (color in online)

A polynomial surface, per stress component (10) (of the same degree as the FE interpolation) is fitted to the FE stress values at the super-convergent points of the patch by using a least square approach, as shown in Fig. 1b:

$$\hat{\sigma}_k^*(\boldsymbol{x}) = \boldsymbol{p}(\boldsymbol{x})\boldsymbol{a}_k \quad k = xx, yy, xy \tag{10}$$

where $\boldsymbol{p}(\boldsymbol{x}) = \{1, x, y\}$ for the linear case and $\boldsymbol{a}_k$ are the corresponding coefficients for each stress component. In this case, each component k of the stress field could be recovered independently by minimizing the following functional:

$$\Phi_{SPR} = \sum_{gp}^{NGP} (\boldsymbol{p}(\boldsymbol{x}_{gp})\boldsymbol{a}_k - \boldsymbol{D}\boldsymbol{\varepsilon}(\boldsymbol{u}^h(\boldsymbol{x}_{gp}))|_k)^2 \tag{11}$$

yielding a linear system of equations per component $\boldsymbol{M}\boldsymbol{a}_k = \boldsymbol{H}_k$, where NGP indicates the number of integration (sampling) points in the patch used for the least squares fitting.

The recovered stresses $\hat{\boldsymbol{\sigma}}_i^*$ at each node are obtained particularizing these surfaces at the patch assembly node with a nodal (continuous) stress representation. Note that this process is computationally efficient as it only requires to solve small systems of equations to obtain the recovered field. The nodal values into each element are interpolated using the FE shape functions N_i, according to (12) and a continuous stress field is obtained, see Fig. 2.

$$\boldsymbol{\sigma}^*(\boldsymbol{x})_{\text{SPR}} = \sum_i N_i(\boldsymbol{x})\hat{\boldsymbol{\sigma}}_i^* \tag{12}$$

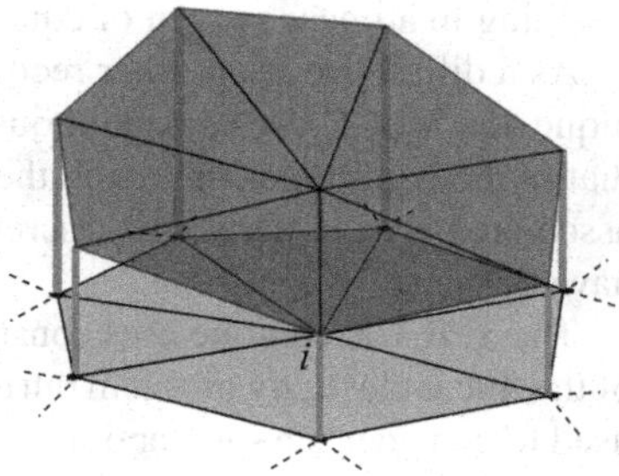

Fig. 2 Representation of the final recovered stress field σ^* over the problem domain. The nodal recovered values are interpolated by the FE shape functions

where N_i is the shape function associated to $i \in \mathcal{N}$ and $\hat{\sigma}_i^*$ is the corresponding recovered nodal stress value.[1]

2.2 A Nearly Equilibrated Recovery Procedure. The SPR-C Technique

Since the introduction of the original SPR technique [4] several contributions aimed at improving the quality and the robustness of this technique. In general, in one way or another, they consider the equations that are satisfied by the exact stress solution in order to improve the quality of the recovered field. Recently, Ródenas and co-workers introduced the so-called SPR-C technique [12] (where the "C" stands for constraints) that was later applied in the XFEM context by Ródenas et al. [13] and finally adapted to geometry-mesh independent FE formulations [14]. As in the SPR technique, a patch $\mathscr{P}^i$ is defined as the set of elements connected to a vertex node i. On each patch, a polynomial expansion for each one of the components of the recovered stress field is expressed according to (10). All the components of the stress vector are simultaneously consider to be able to include the required constrain equations. Thus, in the SPR-C, the recovered stress field for the 2D case, for each patch, reads:

$$\hat{\sigma}_i^*(\boldsymbol{x}) = \begin{Bmatrix} \hat{\sigma}_{xx}^*(\boldsymbol{x}) \\ \hat{\sigma}_{yy}^*(\boldsymbol{x}) \\ \hat{\sigma}_{xy}^*(\boldsymbol{x}) \end{Bmatrix} = \boldsymbol{P}(\boldsymbol{x})\boldsymbol{A} = \begin{bmatrix} \boldsymbol{p}(\boldsymbol{x}) & \boldsymbol{0} & \boldsymbol{0} \\ \boldsymbol{0} & \boldsymbol{p}(\boldsymbol{x}) & \boldsymbol{0} \\ \boldsymbol{0} & \boldsymbol{0} & \boldsymbol{p}(\boldsymbol{x}) \end{bmatrix} \begin{Bmatrix} \boldsymbol{a}_{xx} \\ \boldsymbol{a}_{yy} \\ \boldsymbol{a}_{xy} \end{Bmatrix} \tag{13}$$

To obtain the stress field coefficients $\boldsymbol{A}$, the following functional is minimized:

$$\Phi'(\boldsymbol{A}) := \int_{\mathscr{P}^i} (\boldsymbol{P}(\boldsymbol{x})\boldsymbol{A} - \boldsymbol{D}\boldsymbol{\varepsilon}(\boldsymbol{u}^h(\boldsymbol{x})))^2 \, \mathrm{d}\Omega \tag{14}$$

[1]The recovered field σ^*_{SPR}, obtained using the high quality values of stresses evaluated at the superconvergent integration points, has a convergence rate $p + q$ $(q > 0)$, being p the order to the FE interpolation [11]. This property leads to asymptotically exact error estimators.

resulting in a linear system of equations to solve at each patch $\mathscr{P}^i$.

As a difference from other recovery techniques based on the standard SPR technique, the SPR-C uses a continuous least squares approach. The main implication is that in the continuous approach the values at sampling points are weighted by their associated area, whereas in discrete approach (standard SPR) all sampling points have the same weight.[2]

The SPR-C technique uses constrain equations to consider the known information of the linear elasticity problem during the recovery process. Lagrange multipliers are used to consider the satisfaction of the internal equilibrium equation (int), boundary equilibrium equation (ext) and compatibility equation (cmp), when evaluating the coefficients $\boldsymbol{A}$. The constrain equations to be considered are described below.

- Internal equilibrium equation: the constraint equation for the internal equilibrium in the patch is defined as:

$$c^{\text{int}}(\boldsymbol{x}_j) : \boldsymbol{L}^T \hat{\boldsymbol{\sigma}}_i^*(\boldsymbol{x}_j) + \boldsymbol{b}(\boldsymbol{x}_j) = 0 \qquad \forall \boldsymbol{x}_j \in \mathscr{P}^i \tag{15}$$

  $c^{\text{int}}(\boldsymbol{x}_j)$ is enforced in a sufficient number of non-aligned points ($niee$) to guarantee the exact representation of $\boldsymbol{b}(\boldsymbol{x})$.
- Boundary equilibrium equation: the constraint equation reads:

$$c^{\text{ext}}(\boldsymbol{x}_j) : \boldsymbol{G}\hat{\boldsymbol{\sigma}}_i^*(\boldsymbol{x}_j) = \boldsymbol{t}(\boldsymbol{x}_j) \qquad \forall \boldsymbol{x}_j \in \Gamma_N \cap \mathscr{P}^i \tag{16}$$

  $c^{\text{ext}}(\boldsymbol{x}_j)$ is enforced in $p+1 = nbee$ points along $\Gamma_N \cap \mathscr{P}^i$. In the case where more than one boundary is intersecting the patch, only one curve is considered in order to avoid over-constraining the system of equations.
- Compatibility equation: $\boldsymbol{c}^{\text{cmp}}(\boldsymbol{x}_j)$ is only imposed when $p \geq 2$ in a sufficient number of non-aligned points nc. For example, for $p = 2$ we have $nc = 1$. $\hat{\boldsymbol{\sigma}}^*$ directly satisfies c^{cmp} for $p = 1$. The 2D compatibility equation expressed in terms of stresses, (see [15]) is:

$$c^{\text{cmp}}(\boldsymbol{x}_j) : \frac{\partial^2}{\partial y^2}\left(\kappa\hat{\sigma}_{xx}(\boldsymbol{x}_j) - \nu\delta\hat{\sigma}_{yy}(\boldsymbol{x}_j)\right) + \frac{\partial^2}{\partial x^2}\left(\kappa\hat{\sigma}_{yy}(\boldsymbol{x}_j) - \nu\delta\hat{\sigma}_{xx}(\boldsymbol{x}_j)\right)$$
$$-2(1+\nu)\frac{\partial^2\hat{\sigma}_{xy}(\boldsymbol{x}_j)}{\partial x \partial y} = 0 \qquad \forall \boldsymbol{x}_j \in \mathscr{P}^i \tag{17}$$

  where κ, δ are functions of the Poisson's coefficient ν

$$\begin{cases} \kappa = 1,\ \delta = 1 & \text{for plane stress} \\ \kappa = (1-\nu)^2,\ \delta = (1+\nu) & \text{for plane strain} \end{cases}$$

[2]This is important when the distribution of integration points in the mesh (used as sampling points in the least squares fitting) is not uniform, like in XFEM.

Thus, the functional to be optimized considering the constraint equations for a patch $\mathscr{P}^i$ can be written as:

$$\Phi(\boldsymbol{A}, \boldsymbol{\lambda}) := \Phi'(\boldsymbol{A}) + \sum_{j}^{nbee} \lambda_j^{\mathrm{int}} \left(c^{\mathrm{int}}(\boldsymbol{x}_j)\right) + \sum_{j}^{niee} \lambda_j^{\mathrm{ext}} \left(c^{\mathrm{ext}}(\boldsymbol{x}_j)\right) + \sum_{j}^{nce} \lambda_j^{\mathrm{cmp}} \left(c^{\mathrm{cmp}}(\boldsymbol{x}_j)\right) \quad (18)$$

yielding to a system of equations to be solved at each path to obtain $\hat{\boldsymbol{\sigma}}_i^*(\boldsymbol{x})$.

To obtain a continuous field, a partition of unity procedure (the conjoint polynomial enhancement [16]) properly weighting the stress interpolation polynomials, obtained from patches corresponding to each of the vertex nodes of the element containing point $\boldsymbol{x}$, is used. The field $\boldsymbol{\sigma}_{\mathrm{C}}^*$ is interpolated using linear shape functions N_i associated with the n_v vertex nodes such that:

$$\boldsymbol{\sigma}_{\mathrm{C}}^*(\boldsymbol{x}) = \sum_{J=1}^{n_v} N_i(\boldsymbol{x}) \hat{\boldsymbol{\sigma}}_i^*(\boldsymbol{x}) \quad (19)$$

2.3 Nearly Equilibrated Displacement Recovery Procedure. The SPR-CD Technique

A third approach to obtain a recovered stress field is to directly improve the solution field and then obtain a recovered stress field. In this section we propose to obtain a recovered displacement field form $\boldsymbol{u}^h$, instead of a recovered stress field as in the SPR-C technique. The proposed method, denoted as SPR-CD (where C stands for constraints and D for displacements), is more complete in the sense that it is able to provide an improved recovered pair $(\boldsymbol{u}_u^*, \boldsymbol{\sigma}_\sigma^*)$.[3] The SPR-C technique consists of a displacement recovery procedure where, at each patch, we will be able to impose the satisfaction of the Dirichlet boundary conditions, internal equilibrium equation and boundary equilibrium equation. Figure 3 shows the patch of elements corresponding to the node i with an internal boundary intersecting the patch. This is a more general case where the patch is subdivided into two zones by an internal surface, that could represent the boundary between two materials or a crack surface.

As in the case of SPR-C, SPR-CD is based in subdividing the domain in small regions or patches $\mathscr{P}^i$. In Fig. 3 we show an example of a patch of elements around an assembly node. In the SPR-CD we fit a polynomial surface to each component of the displacement field as we did in [12] for each component of the stress field. Next, we add extra information to improve the solution at each patch. This information is related to the internal and boundary equilibrium and Dirichlet boundary conditions.

[3] Hereinafter subindex u will refer to kinematically admissible global fields and subindex σ will refer to nearly-statically admissible global fields.

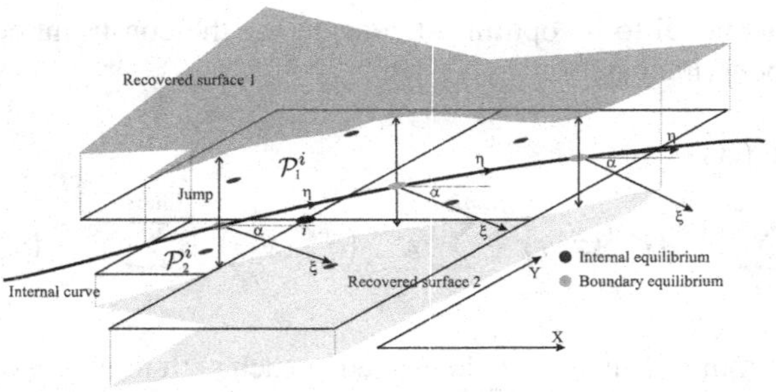

Fig. 3 Example of an internal patch split by an *internal curve*. The *black node* indicates the patch assembly node. The *red points* indicate the position where boundary equilibrium is imposed and the *green points* indicate some random points where the internal equilibrium constraints are imposed. Note that the internal boundary does not necessary follow the element sides. (color in online)

Then a local recovered displacement field $\hat{\boldsymbol{u}}_i^*$ will be evaluated at each patch around the vertex nodes i. $\hat{\boldsymbol{u}}_i^*$ will be evaluated using a scheme similar to that used for the SPR technique [3] but in this case we use the FE displacements of the elements within the patch instead of the corresponding FE stresses. The polynomial surface $\hat{\boldsymbol{u}}_i^*$ will be forced to satisfy the Dirichlet boundary conditions and its corresponding stresses $\hat{\boldsymbol{\sigma}}_i^*(\hat{\boldsymbol{u}}_i^*)$, will be forced to satisfy the internal equilibrium and the boundary equilibrium equations. Note that as $\hat{\boldsymbol{\sigma}}_i^*$ is directly evaluated from $\hat{\boldsymbol{u}}_i^*$, the compatibility equation is satisfied and does not need to be explicitly considered as in the case of the SPR-C technique. The satisfaction of these equations will be enforced using the Lagrange multipliers technique following a point collocation approach in a sufficient number of points, according to the degree of the recovered displacement field. Thus, we define the local recovered displacement and stress fields as follows:

$$\hat{\boldsymbol{u}}_i^*(\boldsymbol{x}) = \begin{Bmatrix} \hat{u}_x^*(\boldsymbol{x}) \\ \hat{u}_y^*(\boldsymbol{x}) \end{Bmatrix} = \boldsymbol{P}(\boldsymbol{x})\boldsymbol{A} = \begin{bmatrix} \boldsymbol{p}(\boldsymbol{x}) & \boldsymbol{0} \\ \boldsymbol{0} & \boldsymbol{p}(\boldsymbol{x}) \end{bmatrix} \begin{Bmatrix} \boldsymbol{a}_x \\ \boldsymbol{a}_y \end{Bmatrix} \tag{20}$$

$$\hat{\boldsymbol{\sigma}}_i^*(\boldsymbol{x}) = \boldsymbol{D}\boldsymbol{L}\hat{\boldsymbol{u}}_i^*(\boldsymbol{x}) \tag{21}$$

Note that, in this case, the degree p of the recovered displacement is one order higher than the degree of the FE nodal interpolation. As in the SPR-C technique, the recovered field is enhanced with known information about boundary conditions and equilibrium to increase the accuracy of the recovered pair $(\boldsymbol{u}_u^*, \boldsymbol{\sigma}_\sigma^*)$. Under the definitions in (20) and (21), the functional to be optimized at each patch $\mathscr{P}^i$ reads as follows:

$$\Phi_D(\boldsymbol{A}, \boldsymbol{\lambda}) := \int_{\mathscr{P}^i} (\boldsymbol{P}(\boldsymbol{x})\boldsymbol{A} - \boldsymbol{u}^h(\boldsymbol{x}))^2 \, \mathrm{d}\Omega +$$

$$\sum_j^{nbee} \lambda_j^{nbee} \left(c^{\mathrm{int}}(\boldsymbol{x}_j)\right) + \sum_j^{niee} \lambda_j^{niee} \left(c^{\mathrm{ext}}(\boldsymbol{x}_j)\right) + \sum_j^{ndce} \lambda_j^{ndce} \left(c^{\mathrm{dir}}(\boldsymbol{x}_j)\right) \tag{22}$$

where $c^{\mathrm{int}}(\boldsymbol{x}_j)$ is defined in (15) and $c^{\mathrm{ext}}(\boldsymbol{x}_j)$ in (16), considering that $\hat{\boldsymbol{\sigma}}_i^*(\boldsymbol{x}_j)$ is defined in (21), and $c^{\mathrm{dir}}(\boldsymbol{x}_j)$ in (23).

- Dirichlet constraints: constraints related with the Dirichlet boundary conditions can be written as:

$$c^{\mathrm{dir}}(\boldsymbol{x}_j) : \hat{\boldsymbol{u}}_i^*(\boldsymbol{x}_j) - \boldsymbol{u}(\boldsymbol{x}_j) = 0 \qquad \forall \boldsymbol{x}_j \in \Gamma_D \cap \mathscr{P}^i \tag{23}$$

We impose the satisfaction of the Dirichlet boundary conditions at $ndce = p + 1$ points (being p the degree of the recovered displacement field) along the part of Γ_D falling into the patch.

The global kinematically admissible displacement field is evaluated at each element using the "Conjoint Polynomial Enhancement" [16], using the displacement field $\hat{\boldsymbol{u}}_i^*$ evaluated from the patches corresponding to each of the n_v vertex nodes of the element and the linear shape functions N_i associated to these nodes:

$$\boldsymbol{u}_u^*(\boldsymbol{x}) = \sum_{i=1}^{n_v} N_i(\boldsymbol{x})\hat{\boldsymbol{u}}_i^*(\boldsymbol{x}) \tag{24}$$

Note that because of the use of (24) we will lose the internal equilibration of the patch recovered stress field $\hat{\boldsymbol{\sigma}}_i^*$ as detailed in Sect. 2.3.1.

2.3.1 Recovered Stress Evaluation

$\boldsymbol{u}_u^*(\boldsymbol{x})$ is a kinematically admissible recovered displacement field. To obtain a consistent recovered stress field we should differentiate (24) according to the following expression:

$$\boldsymbol{\sigma}_u^*(\boldsymbol{x}) = \boldsymbol{D}\boldsymbol{L}\sum_{i=1}^{n_v} N_i(\boldsymbol{x})\hat{\boldsymbol{u}}_i^*(\boldsymbol{x}) = \underbrace{\sum_{i=1}^{n_v} \boldsymbol{D}\left(\boldsymbol{L}N_i(\boldsymbol{x})\right)\hat{\boldsymbol{u}}_i^*(\boldsymbol{x})}_{discontinuous} + \underbrace{\sum_{i=1}^{n_v} N_i(\boldsymbol{x})\,\underbrace{\boldsymbol{D}\boldsymbol{L}\hat{\boldsymbol{u}}_i^*(\boldsymbol{x})}_{\hat{\boldsymbol{\sigma}}_i^*(\boldsymbol{x})}}_{\boldsymbol{\sigma}_\sigma^*(\boldsymbol{x})\ continuous} \tag{25}$$

When we apply the differential operator $\boldsymbol{L}$ to the kinematically admissible displacement field $\boldsymbol{u}_u^*$ we generate the kinematically admissible pair $(\boldsymbol{u}_u^*, \boldsymbol{\sigma}_u^*)$. If we observe equation (25), the field $\boldsymbol{\sigma}_u^*$ is split into two parts, one continuous and one discontinuous. The continuous part coincides with the partition of unity of the patch-wise recovered stress field $\hat{\boldsymbol{\sigma}}_i^*$, directly derived from $\hat{\boldsymbol{u}}_i^*$ (see (21)). Note that $\hat{\boldsymbol{\sigma}}_i^*$ satisfies the equilibrium equations.

It can be said that the main part of the stress field description will be taken into account in the continuous part $\boldsymbol{\sigma}_\sigma^*$, since the discontinuous part will tend to zero with the mesh refinement. Note that in the infinite dimensional space, $\hat{\boldsymbol{u}}_i^*$ will have the same value at each point when evaluated from the different patches and the discontinuous part will be zero because of the partition of nullity of the derivatives of the shape functions. Moreover, the statically admissible stress fields have, in general, a better quality than the kinematically admissible ones since the equilibrium conditions are strongly enforced, and in this particular case we could have a high control of the statical admissibility properties of the continuous part. Because of these reasons and in order to retain the continuity of the recovered stress field we will use only as the continuous part:

$$\boldsymbol{\sigma}_{\sigma}^{*}(\boldsymbol{x})=\sum_{i=1}^{n_v} N_i(\boldsymbol{x})\boldsymbol{D}\boldsymbol{L}\boldsymbol{u}_i^{*}(\boldsymbol{x})=\sum_{i=1}^{n_v} N_i(\boldsymbol{x})\boldsymbol{\sigma}_i^{*}(\boldsymbol{x}) \tag{26}$$

The standard output of the FE code can be the pair $(\boldsymbol{u}_u^*, \boldsymbol{\sigma}_\sigma^*)$ instead of $(\boldsymbol{u}^h, \boldsymbol{\sigma}^h)$. Error estimator for the recovered solution and h-adaptive processes based on the error of the recovered solution will be presented in Sect. 4. Note that, because of the use of the partition of unity technique in (26), $\boldsymbol{\sigma}_\sigma^*$ will not satisfy the internal equilibrium equation:

$$\boldsymbol{L}^T\boldsymbol{\sigma}_{\sigma}^{*}(\boldsymbol{x})=\boldsymbol{L}^T\sum_{i=1}^{n_v} N_i(\boldsymbol{x})\boldsymbol{\sigma}_i^{*}(\boldsymbol{x})=\underbrace{\sum_{i=1}^{n_v}\boldsymbol{L}^T N_i(\boldsymbol{x})\boldsymbol{\sigma}_i^{*}(\boldsymbol{x})}_{-\boldsymbol{s}_\sigma^*}+\underbrace{\sum_{i=1}^{n_v} N_i(\boldsymbol{x})\boldsymbol{L}^T\boldsymbol{\sigma}_i^{*}(\boldsymbol{x})}_{-\boldsymbol{b}} \tag{27}$$

This expression is a modified version of the internal equilibrium equation where $\boldsymbol{s}_\sigma^*$ represents the lack of internal equilibrium. Furthermore, there could also exist a lack of boundary equilibrium of $\boldsymbol{\sigma}_\sigma^*$ over Γ_N which can be evaluated as $\boldsymbol{r}_\sigma^* = \boldsymbol{G}\boldsymbol{\sigma}_\sigma^* - \boldsymbol{t}_{\Gamma_N}$, where $\boldsymbol{t}_{\Gamma_N}$ are the exact tractions over the Neumann boundaries.

2.4 Singular Fields

The SPR-CD recovery technique can also consider the singular behaviour of the solution due to a re-entrant corner, a crack, etc. We have to take into account that the SPR-based techniques tend to increase the smoothness of the solution. However this characteristic, being convenient for standard situations, can decrease the accuracy in the surroundings of the singularities. Moreover, in the XFEM framework the recovered field requires to be enriched with the asymptotic fields of the singular solution to obtain satisfactory results [17]. Different techniques have been proposed to account for the singular behaviour during the recovery process [18–20]. Here, following the ideas in [18], for stresses in singular problems, our solution $(\boldsymbol{u}, \boldsymbol{\sigma})$ will be split into 2 parts, one singular $(\boldsymbol{u}_{\text{sing}}, \boldsymbol{\sigma}_{\text{sing}})$ and one smooth $(\boldsymbol{u}_{\text{smo}}, \boldsymbol{\sigma}_{\text{smo}})$:

$$\begin{aligned}\boldsymbol{u}(\boldsymbol{x}) &= \boldsymbol{u}_{\text{smo}}(\boldsymbol{x}) + \boldsymbol{u}_{\text{sing}}(\boldsymbol{x})\\ \boldsymbol{\sigma}(\boldsymbol{x}) &= \boldsymbol{\sigma}_{\text{smo}}(\boldsymbol{x}) + \boldsymbol{\sigma}_{\text{sing}}(\boldsymbol{x})\end{aligned} \tag{28}$$

Note that the displacement field $\boldsymbol{u}$ does not have any singular behaviour, however we maintain the subscripts "smo" and "sing" for consistency. The recovered fields $(\boldsymbol{u}^*_u, \boldsymbol{\sigma}^*_\sigma)$ can also be expressed as the contribution of two recovered fields, one smooth and one singular:

$$\begin{aligned} \boldsymbol{u}^*_u(\boldsymbol{x}) &= \boldsymbol{u}^*_{\text{smo}}(\boldsymbol{x}) + \boldsymbol{u}^*_{\text{sing}}(\boldsymbol{x}) \\ \boldsymbol{\sigma}^*_\sigma(\boldsymbol{x}) &= \boldsymbol{\sigma}^*_{\text{smo}}(\boldsymbol{x}) + \boldsymbol{\sigma}^*_{\text{sing}}(\boldsymbol{x}) \end{aligned} \tag{29}$$

For the recovery of the singular part we will use the expressions which describe the asymptotic fields near the crack tip with respect to a coordinate system (r, ϕ) at the tip as described in [21]:

$$\begin{aligned} \mathbf{u}_{\text{sing}}(r, \phi) &= K_{\text{I}} r^{\lambda_{\text{I}}} \boldsymbol{\Psi}_{\text{I}}(\lambda_{\text{I}}, \phi) + K_{\text{II}} r^{\lambda_{\text{II}}} \boldsymbol{\Psi}_{\text{II}}(\lambda_{\text{II}}, \phi) \\ \boldsymbol{\sigma}_{\text{sing}}(r, \phi) &= K_{\text{I}} \lambda_{\text{I}} r^{\lambda_{\text{I}}-1} \boldsymbol{\Phi}_{\text{I}}(\lambda_{\text{I}}, \phi) + K_{\text{II}} \lambda_{\text{II}} r^{\lambda_{\text{II}}-1} \boldsymbol{\Phi}_{\text{II}}(\lambda_{\text{II}}, \phi) \end{aligned} \tag{30}$$

where r is the radial distance to the corner, λ_m (with $m = \text{I, II}$) are the eigenvalues that determine the order of the singularity, $\boldsymbol{\Psi}_m$ and $\boldsymbol{\Phi}_m$ are sets of trigonometric functions that depend on the angular position ϕ, and K_m are the so-called Generalized Stress Intensity Factors (GSIFs). The GSIF is a multiplicative constant that depends on the loading of the problem and linearly determines the intensity of the displacement and stress fields in the vicinity of the singular point. Therefore, the eigenvalues λ and the Generalized Stress Intensity Factor (GSIF) K define the singular field. The Generalized Stress Intensity Factor K is the characterizing parameter in fracture mechanics problems with singularities. In the particular case in which $2\alpha = 360°$ the problem will correspond to a crack as considered in the context of Linear Elastic Fracture Mechanics where this parameter is called the Stress Intensity Factor (SIF). An accurate value of K can be estimated from the FE solution using different techniques like the interaction integral. For further information see [21–23] and the references therein.

Once K has been obtained, the singular field $(\boldsymbol{u}^*_{\text{sing}}, \boldsymbol{\sigma}^*_{\text{sing}})$ will be evaluated as in (30). $\boldsymbol{u}^h_{\text{smo}}$ is defined as the result of subtracting the singular part $\boldsymbol{u}^*_{\text{sing}}$ from the FE solution $\boldsymbol{u}^h$:

$$\boldsymbol{u}^h_{\text{smo}}(\boldsymbol{x}) = \boldsymbol{u}^h(\boldsymbol{x}) - \boldsymbol{u}^*_{\text{sing}}(\boldsymbol{x}) \tag{31}$$

$\boldsymbol{u}^h_{\text{smo}}$ will be used as the input for the SPR-CD recovery process that will yield the pair $(\boldsymbol{u}^*_{\text{smo}}, \boldsymbol{\sigma}^*_{\text{smo}})$. The final recovered solution will be evaluated using (29). Note that $\boldsymbol{L}^T \boldsymbol{\sigma}^*_{\text{sing}} = 0$, therefore no additional terms will be considered in (15) when applying the SPR-CD technique to $\boldsymbol{\sigma}^h_{\text{smo}} = \boldsymbol{\sigma}^h - \boldsymbol{\sigma}^h_{\text{sing}}$. However, (16) will be modified by subtracting the singular part of the traction $\boldsymbol{t}^*_{\text{sing}} = \boldsymbol{G}\boldsymbol{\sigma}^*_{\text{sing}}$ from the r.h.s. of the equation. The same occur with the constrains related to the Dirichlet boundaries, (23), where the r.h.s will be modified by subtracting the singular part of the displacement field $\boldsymbol{u}^*_{\text{sing}}$.

This splitting procedure is not used in the whole domain of the problem but only in an area close to the singular point in order to localize the process allowing for a number of different singularities in the same problem. The area affected by this process is defined by the user, and it should be related with extension of the area dominated by the singular behaviour of the stress field.

3 Error Estimation and Bounding

3.1 Error Estimation in Energy Norm

Recovery-type error estimators relay on the use of the Zienkiewicz and Zhu (ZZ) error estimators introduced in (9). The quality of the estimation is highly related to the quality of the recovered field $\boldsymbol{\sigma}^*$, so the accuracy of the estimate will be strongly affected by the quality of the recovered field. We will compare two different estimators, the first one with the stress field provided by the SPR-C technique $\boldsymbol{\sigma}^*_{\mathrm{C}}$, and the second one with the stress field provided by the SPR-CD technique $\boldsymbol{\sigma}^*_{\sigma}$:

$$\mathscr{E}^2_{\mathrm{C}} := \int_{\Omega} (\boldsymbol{\sigma}^*_{\mathrm{C}} - \boldsymbol{\sigma}^h)^T \boldsymbol{D}^{-1} (\boldsymbol{\sigma}^*_{\mathrm{C}} - \boldsymbol{\sigma}^h) \ \mathrm{d}\Omega \tag{32}$$

$$\mathscr{E}^2_{\mathrm{CD}} := \int_{\Omega} (\boldsymbol{\sigma}^*_{\sigma} - \boldsymbol{\sigma}^h)^T \boldsymbol{D}^{-1} (\boldsymbol{\sigma}^*_{\sigma} - \boldsymbol{\sigma}^h) \ \mathrm{d}\Omega \tag{33}$$

Note that both fields, $\boldsymbol{\sigma}^*_{\mathrm{C}}$ and $\boldsymbol{\sigma}^*_{\sigma}$, are locally equilibrated, but SPR-C is a stress-based smoothing technique, while SPR-CD is displacement-based smoothing technique. We expect similar results, but the SPR-CD recovery process, having a similar computational cost than the SPR-C, will provide us also a recovered displacement field that will be useful in the following sections.

3.2 Error Estimation in Quantities of Interest

In the previous sections we have introduced the ZZ error estimator in energy norm. However, for some applications it is also interesting to measure the error in a different norm. For instance if we are interested in the quality of the displacement field in a certain region of the domain, the energy norm, obviously, is not the best suitable norm. In order to deal with these situations we introduce the error estimator in Quantities of Interest (QoI). The ZZ error estimator can be easy extended to an error estimator in QoI according to the following expressions:

$$\tilde{\mathscr{E}}^2_{\mathrm{C}} := \int_{\Omega} (\tilde{\boldsymbol{\sigma}}^*_{\mathrm{C}} - \tilde{\boldsymbol{\sigma}}^h)^T \boldsymbol{D}^{-1} (\boldsymbol{\sigma}^*_{\mathrm{C}} - \boldsymbol{\sigma}^h)\ \mathrm{d}\Omega \tag{34}$$

$$\tilde{\mathscr{E}}^2_{\mathrm{CD}} := \int_{\Omega} (\tilde{\boldsymbol{\sigma}}^*_{\sigma} - \tilde{\boldsymbol{\sigma}}^h)^T \boldsymbol{D}^{-1} (\boldsymbol{\sigma}^*_{\sigma} - \boldsymbol{\sigma}^h)\ \mathrm{d}\Omega \tag{35}$$

where ˜ indicates that fields refers to the solution of the auxiliary or dual problem. The dual problem is used to extract the required information from the main (or primal) problem we are solving. A full description of this technique can be found in [24] and references therein.

3.3 A SPR-Based Upper Bounding Technique

In this section, we will first introduce a general upper bounding technique of the error in energy norm for general SPR-based recovery procedures. The upper bounding technique presented consists of two parts. The first one is the classical ZZ error estimator, providing accuracy to the error estimator, and the second part is an explicit-type residual-based error estimator to bound the lack of equilibrium of the recovered stress field, providing the required bounding properties. Then, we will show an alternative procedure based on the residual of the recovered solution to obtain this bound. We will apply the results to the particular case in which the recovered solution is obtained with the SPR-CD recovery technique. Finally, we will show a procedure to obtain a numerical value for the constant required for bounding the correction terms when the SPR-CD recovery technique is used.

Thanks to the accuracy of the recovered field evaluated with SPR-type techniques, this asymptotically exact error estimator produces very accurate estimations of the exact error, especially with enhanced versions of the original SPR like the SPR-C or SPR-CD techniques. However, under this framework, we need a statically admissible recovered stress field in order to obtain an upper bound of the error in the energy norm. This means that this stress field has to be in equilibrium with the body loads and the Neumann tractions.

The basic SPR-type techniques produce continuous fields $\boldsymbol{\sigma}^*$ but they fail to obtain a statically admissible stress field. The SPR-based techniques, in general, introduce a lack of internal equilibrium $\boldsymbol{s}^*$ and boundary equilibrium $\boldsymbol{r}^*$, as previously described. Díez et al. [25] introduced an expression to evaluate a recovery-based upper bound of the error in the energy norm with a correction term to account for the lack of internal equilibrium. A generalization of that expression is introduced in (36), which does not only include the lack of internal equilibrium but also the lack of boundary equilibrium making use of the two correction terms shown in (37):

$$|||\boldsymbol{e}|||^2_{\Omega} \leq \mathscr{E}^2_{\mathrm{UB}} := \mathscr{E}^2 + \mathscr{E}_{\mathrm{int}} + \mathscr{E}_{\mathrm{bnd}} \tag{36}$$

$$\mathscr{E}_{\text{int}} := -2 \int_{\Omega} (\boldsymbol{s}^*)^T \boldsymbol{e} \ \mathrm{d}\Omega$$
$$\mathscr{E}_{\text{bnd}} := -2 \int_{\Gamma_N} (\boldsymbol{r}^*)^T \boldsymbol{e} \ \mathrm{d}\Gamma \tag{37}$$

These correction terms require the exact displacement solution, $\boldsymbol{e}$, to be evaluated. In (36) $\mathscr{E}$ can be evaluated using the ZZ error estimator. In this section we are interested in bounding (37) in order to obtain guaranteed upper bounds of the error in the energy norm. These correction terms are first bounded with the Cauchy-Schwarz inequality:

$$\left| \int_{\Omega} (\boldsymbol{s}^*)^T \boldsymbol{e} \ \mathrm{d}\Omega \right| \leq \left\| \boldsymbol{s}^* \right\|_{L_2(\Omega)} \|\boldsymbol{e}\|_{L_2(\Omega)}$$
$$\left| \int_{\Gamma_N} (\boldsymbol{r}^*)^T \boldsymbol{e} \ \mathrm{d}\Gamma \right| \leq \left\| \boldsymbol{r}^* \right\|_{L_2(\Gamma_N)} \|\boldsymbol{e}\|_{L_2(\Gamma_N)} \tag{38}$$

As $\boldsymbol{s}^*$ and $\boldsymbol{r}^*$ can be evaluated, the correction terms in (38) will be bounded if we obtain a bound of the L_2-norm of the error in the displacement field. With the Aubin-Nitsche lemma ([26] p. 136), the L_2-norm of the error in the displacement field can be bounded with the respective error in the energy norm, then:

$$\|\boldsymbol{e}\|_{L_2(\Omega)} \leq C_{\Omega} h \,|||\boldsymbol{e}|||_{\Omega} \tag{39}$$

Additionally, making use of the trace inequality [27] we can also bound the error evaluated in L_2-norm over the boundary with the error in energy norm according to the following expression:

$$\|\boldsymbol{e}\|_{L_2(\Gamma_N)} \leq C_{\Gamma} h^{\frac{1}{2}} \,|||\boldsymbol{e}|||_{\Omega} \tag{40}$$

where C_{Ω} and C_{Γ} are independent of the mesh size, and h is a relative representative size of the mesh (in an h-uniform refinement process h is the element size). It is important to remark that standard residual-based error estimators are usually derived using Clément-type inequalities [27, 28] thanks to the orthogonality between the FE solution and the error. However, in our case the recovered stress field does not satisfy the Galerkin orthogonality.

The inequalities in (39) and (40) were developed for h-uniform refinement processes. However they can be extended for constant-pattern h-adaptive refinement processes, since for a defined mesh pattern it is possible to find a geometrical mapping that converts the h-adapted mesh to an uniform one. The constant-pattern meshes can be found in standard h-adaptive refinement processes when the asymptotic range is achieved. The refinement processes start from a mesh of elements of uniform size. During the process, in the pre-asymptotic range, the h-adaptive technique will adequately refine the zones which need a smaller element size in order to obtain an equidistribution of the discretization error. Once the equidistribution of the error is achieved, in the asymptotic range, the refinement process essentially consists of a uniform refinement of the h-adapted mesh. This can be defined as constant-pattern h-adaptive refinement processes. Under this situation, h can be related with the Number of Degrees of Freedom (NDoF). Therefore we define $h := \left(\frac{1}{NDoF}\right)^{\frac{1}{d}}$, where d is the dimension of the problem. Then, we can bound the correction terms as follows:

$$\begin{aligned} \mathscr{E}_{\text{int}} &\leq 2\left|\int_{\Omega} (\boldsymbol{s}^*)^T \boldsymbol{e}\ \mathrm{d}\Omega\right| \leq 2\left\|\boldsymbol{s}^*\right\|_{L_2(\Omega)} \|\boldsymbol{e}\|_{L_2(\Omega)} \leq C_\Omega h \left\|\boldsymbol{s}^*\right\|_{L_2(\Omega)} |||\boldsymbol{e}|||_\Omega \\ \mathscr{E}_{\text{bnd}} &\leq 2\left|\int_{\Gamma_N} (\boldsymbol{r}^*)^T \boldsymbol{e}\ \mathrm{d}\Gamma\right| \leq 2\left\|\boldsymbol{r}^*\right\|_{L_2(\Gamma_N)} \|\boldsymbol{e}\|_{L_2(\Gamma_N)} \leq C_\Gamma h^{\frac{1}{2}} \left\|\boldsymbol{r}^*\right\|_{L_2(\Gamma_N)} |||\boldsymbol{e}|||_\Omega \end{aligned} \tag{41}$$

3.3.1 Particular Case Using the SPR-C Recovery Technique

Let us now consider the evaluation of the upper bound of the discretization error in the energy norm when the SPR-CD technique is used instead of a standard SPR-based procedure: $\boldsymbol{\sigma}^* = \boldsymbol{\sigma}^*_\sigma$. Because of the use of the SPR-CD technique, the lack of equilibrium along the boundary is either null or negligible, thus we can neglect the terms related to the lack of equilibrium along the boundaries $\left\|\boldsymbol{r}^*_\sigma\right\|_{L_2(\Gamma_N)}$. Then, with minimal loss in accuracy:

$$\mathscr{E}_{\text{int}} + \mathscr{E}_{\text{bnd}} \leq Ch\, |||\boldsymbol{e}|||_\Omega \left\|\boldsymbol{s}^*_\sigma\right\|_{L_2(\Omega)} \tag{42}$$

where $C = 2C_\Omega$. Now, we rewrite expression (36) composed by the ZZ error estimator (9) and the correction terms (37) as follows:

$$|||\boldsymbol{e}|||^2_\Omega \leq \int_\Omega \left(\boldsymbol{\sigma}^*_\sigma - \boldsymbol{\sigma}^h\right)^T \boldsymbol{D}^{-1} \left(\boldsymbol{\sigma}^*_\sigma - \boldsymbol{\sigma}^h\right)\ \mathrm{d}\Omega + \mathscr{E}_{\text{int}} + \mathscr{E}_{\text{bnd}} \tag{43}$$

substituting (42) in (43) we obtain:

$$|||\boldsymbol{e}|||^2_\Omega \leq \mathscr{E}^2_{\text{CD}} + Ch\, |||\boldsymbol{e}|||_\Omega \left\|\boldsymbol{s}^*_\sigma\right\|_{L_2(\Omega)} = \mathscr{E}^2_{\text{CD}} + \Xi\, |||\boldsymbol{e}|||_\Omega \tag{44}$$

Expression (44) is a second order degree polynomial in $|||\boldsymbol{e}|||_\Omega$. The most conservative root provides the upper bound in the energy norm up to the constant C, see [25],

$$|||\boldsymbol{e}|||_\Omega \leq \frac{\Xi + \sqrt{\Xi^2 + 4\mathscr{E}^2_{\text{CD}}}}{2} \tag{45}$$

Now we have to investigate the convergence rate of each term in (44). Assuming that p is the order of the FE interpolation:

- $\mathscr{E}_{CD}$: this term is considered *asymptotically exact* because the recovery technique has a higher convergence rate than the FE solution [5]. Thus its convergence rate could be considered the same as that of the error in the energy norm, p.
- $|||\boldsymbol{e}|||_\Omega$: obviously the convergence rate of this term is p for regular solutions.
- Ξ: ($\Xi = Ch \left\|\boldsymbol{s}_\sigma^*\right\|_{L_2(\Omega)}$) the convergence rate of this term is not totally straightforward. In our case, it will depend on the convergence rate of $\left\|\boldsymbol{s}_\sigma^*\right\|_{L_2(\Omega)}$. Assuming that the recovered field has a convergence rate $p+q$, $q>0$, higher than that for the FE solution, p, the convergence rate of $\left\|\boldsymbol{s}_\sigma^*\right\|_{L_2(\Omega)}$ would be $p+q-1$. Then, under this situation, the convergence rate of Ξ could be considered as $p+q$.

Although we cannot guarantee superconvergence ($q = 1$) we know that the recovered field converges faster than the FE solution, $q > 0$ [11]. This means that the correction terms tend to vanish along the refinement process. Therefore, the plain ZZ error estimator with the SPR-C technique will provide asymptotically guaranteed upper bounds of the error in energy norm.

Finally, we can conclude that provided the recovered field converges faster than the FE solution, we obtain a stable upper bound. With regards to the constant C, some authors take the value $C = 1$ for convergence purposes [27]. However, for error estimation purposes, this approach could be inaccurate in some cases and in other it would provide under estimations of the true error. In the following section we will show a methodology to numerically compute this constant for each problem.

3.3.2 Numerical Evaluation of C_Ω

Expression (45) is an upper bound of the error in the energy norm, but requires the evaluation of a constant C_Ω, which is specific for each problem and also for each discretization type. This constant relates the L_2-norm of the error in displacements with the respective error in the energy norm, as shown in (39).

In this work we propose a methodology to numerically estimate the value of the constant C_Ω to be used in (45) for each problem. We assume that we are under an h-refinement process which is required to evaluate the solution with the level of accuracy defined by the user. Therefore, a set of h-adapted meshes will be available. Let H and h be the representative sizes of two meshes such that $H \ll h$ being $\boldsymbol{u}^H$ the FE solution of the finer mesh. We can consider that $\boldsymbol{u}^H$ is a good approximation to $\boldsymbol{u}$ in comparison with $\boldsymbol{u}^h$. Considering the Richardson extrapolation the following relations will hold:

$$\begin{aligned}
\left\|\boldsymbol{u}-\boldsymbol{u}^h\right\|^2_{L_2(\Omega)} &\approx \frac{\left\|\boldsymbol{u}^H-\boldsymbol{u}^h\right\|^2_{L_2(\Omega)}}{1+\left(\frac{H}{h}\right)^{2p+2}} \\
\left|\left|\left|\boldsymbol{u}-\boldsymbol{u}^h\right|\right|\right|^2_{\Omega} &\approx \frac{\left|\left|\left|\boldsymbol{u}^H-\boldsymbol{u}^h\right|\right|\right|^2_{\Omega}}{1+\left(\frac{H}{h}\right)^{2p}}
\end{aligned} \tag{46}$$

where p is the degree of the FE solution. In order to keep the information of the solution in the finer mesh, the integration will be performed in the finer mesh. Under these assumptions, it easily follows the evaluation of the numerical approximation C^*_Ω to C_Ω for the mesh H.

$$C_\Omega \approx C^*_\Omega = \sqrt{\frac{\|\boldsymbol{u}^H - \boldsymbol{u}^h\|^2_{L_2(\Omega)} \left(1 + \left(\frac{H}{h}\right)^{2p}\right)}{h^2 \,|||\boldsymbol{u}^H - \boldsymbol{u}^h|||^2_\Omega \left(1 + \left(\frac{H}{h}\right)^{2p+2}\right)}} \tag{47}$$

Then the constant C will be also approximated by $C^* = C^*_\Omega$. For other recovery processes, where the boundary equilibrium is not fulfilled, such as in the case of the plain SPR, this cannot be assumed. The numerical evaluation of C can be performed during the refinement process for each mesh n, using information from previous meshes. For the first one, since no previous information is available, it will be considered that $C = 1$. For the second mesh, the constant will be evaluated comparing the solutions of the first $\boldsymbol{u}^1$ and second $\boldsymbol{u}^2$ meshes. Therefore, $\boldsymbol{u}^H = \boldsymbol{u}^2$ and $\boldsymbol{u}^h = \boldsymbol{u}^1$. From the third mesh, $n > 2$, we take, in general, $\boldsymbol{u}^H = \boldsymbol{u}^n$ and $\boldsymbol{u}^h = \boldsymbol{u}^{n-2}$. Note that this process will only be required for the initial meshes of the h-adaptive process until a stable value of the constant is achieved.

Finally the upper bound of the error in the energy norm reads:

$$|||\boldsymbol{e}|||^2_\Omega \lesssim \hat{\mathscr{E}}_{\mathrm{UB}} := \frac{\Xi^* + \sqrt{(\Xi^*)^2 + 4\mathscr{E}^2}}{2} \tag{48}$$

where:

$$\Xi^* = C^* h \left\|\boldsymbol{s}^*_\sigma\right\|^2_{L_2(\Omega)} \tag{49}$$

3.4 Error Lower Bounds in Energy Norm

The next step is to obtain a lower bound on the error in energy norm by using the recovered fields at hand. Taking the ideas presented in [29] it is possible to obtain a lower bound of this value from a nearly-statically admissible solution in combination with a kinematically admissible solution. The proof of the following theorem can be found in [23].

Theorem 1 *Being $\boldsymbol{u}$ the exact displacement solution of the linear elasticity problem presented in Sect. 1, $\boldsymbol{u}^h$ the corresponding approximation obtained with a displacement-based FEM, $\boldsymbol{u}^*_u$ a kinematically admissible displacement field, $\boldsymbol{\sigma}^*_\sigma$ a nearly-statically admissible stress field with $\boldsymbol{s}^*_\sigma$ and $\boldsymbol{r}^*_\sigma$ the corresponding lacks in internal and boundary equilibrium, respectively. The following expression*

$$2\lambda\left\{\bar{a}(\boldsymbol{\sigma}_{\sigma}^{*}-\boldsymbol{\sigma}^{h},\sigma(\boldsymbol{e}_{u}))-\int_{\Omega}(\boldsymbol{s}_{\sigma}^{*})^{T}\boldsymbol{e}_{u}\ \mathrm{d}\Omega-\int_{\Gamma}(\boldsymbol{r}_{\sigma}^{*})^{T}\boldsymbol{e}_{u}\ \mathrm{d}\Gamma\right\}-\lambda^{2}\,|||\boldsymbol{e}_{u}|||_{\Omega}^{2} \tag{50}$$

is a lower error bound in energy norm for any $\lambda \in \mathbb{R}$.

The previous expression is valid for any $\lambda \in \mathbb{R}$. Differentiating with respect to λ we obtain the expression for the optimum λ and thus, the optimum lower bound.

$$\mathscr{E}_{LB}^{2}=\frac{\left\{a(\boldsymbol{e}_{\sigma},\boldsymbol{e}_{u})-\int_{\Omega}(\boldsymbol{s}_{\sigma}^{*})^{T}\boldsymbol{e}_{u}\ \mathrm{d}\Omega-\int_{\Gamma}(\boldsymbol{r}_{\sigma}^{*})^{T}\boldsymbol{e}_{u}\ \mathrm{d}\Gamma\right\}^{2}}{|||\boldsymbol{e}_{u}|||_{\Omega}^{2}}\leq|||\boldsymbol{e}|||_{\Omega}^{2} \tag{51}$$

4 Error Estimation in the Recovered Solution Field

For the recovery-based error estimators we usually compute an enhanced stress solution which is compared with the raw FE stress solution. As explained in Sect. 1 the enhanced solution has better properties, in terms of continuity and equilibrium. However, in order to use it as the output of our FE code we need a method to assess the quality of this magnitude, i.e., we need to obtain an error estimate for the enhanced solution. Some authors [30] evaluate the error of the plain SPR recovered field by comparing it with some enhancements of the SPR solution, such as adding equilibrium and information of the elasticity problem. Other authors [31] also use a recovered (recycled) solution from the recovered solution.

This section will show a technique to estimate the error, for the enhanced SPR-CD technique presented in Sect. 2.3. Recovery error estimation techniques are based on the assumption that the recovered solution is more accurate than the FE solution. A sufficiently accurate estimation of the error in energy norm of the recovered solution could lead to refinement processes based on the accuracy of the recovered solution instead of that of the FE solution. This could result in a considerable reduction of the computational cost that would be particularly interesting, for example, in optimization processes whose efficiency would be significantly increased as the required level of accuracy would be achieved with a considerable lower number of degrees of freedom.

Figure 4 shows an scheme of the evolution of the exact errors in energy norm of the finite element and recovered solutions. Note that in the SPR-CD technique the recovered solution is in practice more accurate than the FE solution and has a higher convergence rate. Therefore, the number of degrees of freedom N_B required by the recovered solution to reach a prescribed accuracy level defined by the blue horizontal line in Fig. 4 is considerably smaller than that required by the FE solution (N_C). We can thus define a highly efficient h-adaptive refinement process that considers the accuracy of the recovered solution instead of the accuracy of the FE solution so that the refinement process finishes when the error in the recovered solution is smaller than the prescribed value.

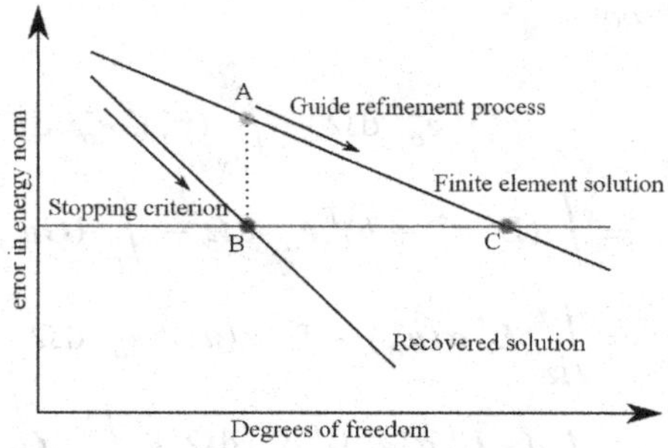

Fig. 4 *h*-adaptive refinement scheme

This section will first show the initial developments that finally led to an heuristic expression that can be used to efficiently estimate the error in energy norm of the field $\boldsymbol{\sigma}_{\sigma}^{*}$, initially evaluated to estimate the error of the FE solution $\boldsymbol{u}^{h}$. The results will show a good accuracy in the error estimation and also an important reduction in computational cost in comparison with standard h-adaptive refinement procedures based on the error in energy norm of the FE solution.

4.1 Error Norm Representation for the Recovered Solution

The purpose of this section is to introduce an error measure for the SPR-CD stress recovered field. We have developed an expression which allows us to evaluate the error of the nearly-statically recovered stress field $\boldsymbol{\sigma}_{\sigma}^{*}$.

The SPR-CD procedure yields a post-processed stress field $\boldsymbol{\sigma}_{\sigma}^{*}$ which is taken as an enhanced approximation to the exact stresses, $\boldsymbol{\sigma}$, more accurate than $\boldsymbol{\sigma}^{h}$. As explained in Sect. 2.3, the recovered stress $\boldsymbol{\sigma}_{\sigma}^{*}$ is continuous but it has a lack of internal equilibrium $\boldsymbol{s}_{\sigma}^{*}$ and also a lack of boundary equilibrium $\boldsymbol{r}_{\sigma}^{*}$.

Theorem 2 *Under the assumptions presented so far, the following expression evaluates the error in energy norm of the recovered stress field.*

$$\begin{aligned} |||\boldsymbol{e}_{\sigma}^{*}|||_{\Omega}^{2} &= |||\boldsymbol{u}-\boldsymbol{u}_{\sigma}^{*}|||_{\Omega}^{2} = \bar{a}\left(\boldsymbol{\sigma}-\boldsymbol{\sigma}_{\sigma}^{*}, \boldsymbol{\sigma}-\boldsymbol{\sigma}_{\sigma}^{*}\right) \\ &= -\int_{\Omega} (\boldsymbol{s}_{\sigma}^{*})^{T} \boldsymbol{e}_{\sigma}^{*} \ \mathrm{d}\Omega - \int_{\Gamma_N} (\boldsymbol{r}_{\sigma}^{*})^{T} \boldsymbol{e}_{\sigma}^{*} \ \mathrm{d}\Gamma \end{aligned} \tag{52}$$

being $\boldsymbol{e}_{\sigma}^{*}$ the error in displacements corresponding to the recovered solution $\boldsymbol{\sigma}_{\sigma}^{*}$ such that $\boldsymbol{\sigma}_{\sigma}^{*} = \boldsymbol{\sigma}(\boldsymbol{u}_{\sigma}^{*}) = \boldsymbol{\sigma}(\boldsymbol{u}-\boldsymbol{e}_{\sigma}^{*})$. Note that $\boldsymbol{u}_{\sigma}^{*}$ and $\boldsymbol{e}_{\sigma}^{*}$ will not be explicitly evaluated.

Proof

$$
\begin{aligned}
&-\int_{\Omega}(\boldsymbol{s}_{\sigma}^{*})^{T}\boldsymbol{e}_{\sigma}^{*}\,\mathrm{d}\Omega-\int_{\Gamma_N}(\boldsymbol{r}_{\sigma}^{*})^{T}\boldsymbol{e}_{\sigma}^{*}\,\mathrm{d}\Gamma\\
&=\int_{\Omega}(\boldsymbol{L}^{T}\boldsymbol{\sigma}_{\sigma}^{*}+\boldsymbol{b})^{T}\boldsymbol{e}_{\sigma}^{*}\,\mathrm{d}\Omega-\int_{\Gamma_N}(\boldsymbol{G}\boldsymbol{\sigma}_{\sigma}^{*}-t)^{T}\boldsymbol{e}_{\sigma}^{*}\,\mathrm{d}\Gamma\\
&=\int_{\Omega}(\boldsymbol{L}^{T}\boldsymbol{\sigma}(\boldsymbol{u}_{\sigma}^{*})-\boldsymbol{L}^{T}\boldsymbol{\sigma}(\boldsymbol{u}))^{T}\boldsymbol{e}_{\sigma}^{*}\,\mathrm{d}\Omega-\int_{\Gamma_N}(\boldsymbol{G}\boldsymbol{\sigma}(\boldsymbol{u}_{\sigma}^{*})-\boldsymbol{G}\boldsymbol{\sigma}(\boldsymbol{u}))^{T}\boldsymbol{e}_{\sigma}^{*}\,\mathrm{d}\Gamma\\
&=\int_{\Omega}(-\boldsymbol{L}^{T}\boldsymbol{\sigma}(\boldsymbol{e}_{\sigma}^{*}))^{T}\boldsymbol{e}_{\sigma}^{*}\,\mathrm{d}\Omega+\int_{\Gamma_N}(\boldsymbol{G}\boldsymbol{\sigma}(\boldsymbol{e}_{\sigma}^{*}))^{T}\boldsymbol{e}_{\sigma}^{*}\,\mathrm{d}\Gamma\\
&=\int_{\Omega}\boldsymbol{\sigma}(\boldsymbol{e}_{\sigma}^{*})^{T}\boldsymbol{\varepsilon}(\boldsymbol{e}_{\sigma}^{*})\,\mathrm{d}\Omega-\int_{\Gamma_N}(\boldsymbol{G}\boldsymbol{\sigma}(\boldsymbol{e}_{\sigma}^{*}))^{T}\boldsymbol{e}_{\sigma}^{*}\,\mathrm{d}\Gamma+\int_{\Gamma_N}(\boldsymbol{G}\boldsymbol{\sigma}(\boldsymbol{e}_{\sigma}^{*}))^{T}\boldsymbol{e}_{\sigma}^{*}\,\mathrm{d}\Gamma\\
&=\int_{\Omega}\boldsymbol{\sigma}(\boldsymbol{e}_{\sigma}^{*})^{T}\boldsymbol{\varepsilon}(\boldsymbol{e}_{\sigma}^{*})\,\mathrm{d}\Omega\\
&=|||\boldsymbol{e}_{\sigma}^{*}|||_{\Omega}^{2}
\end{aligned}
\tag{53}
$$

In the following we will show how to evaluate an upper error bound of $|||\boldsymbol{e}_{\sigma}^{*}|||_{\Omega}^{2}$ and also several heuristic error estimators making use of the recovered solution already evaluated with the SPR-CD technique. Regarding the evaluation of an upper bound, operating with (52) considering Cauchy-Schwarz inequality we will have:

$$
\begin{aligned}
|||\boldsymbol{e}_{\sigma}^{*}|||_{\Omega}^{2}&=-\int_{\Omega}(\boldsymbol{s}_{\sigma}^{*})^{T}\boldsymbol{e}_{\sigma}^{*}\,\mathrm{d}\Omega-\int_{\Gamma_N}(\boldsymbol{r}_{\sigma}^{*})^{T}\boldsymbol{e}_{\sigma}^{*}\,\mathrm{d}\Gamma\\
&\leq\left(\left\|\boldsymbol{s}_{\sigma}^{*}\right\|_{L^2(\Omega)}+\left\|\boldsymbol{r}_{\sigma}^{*}\right\|_{L^2(\Omega)}\right)\left\|\boldsymbol{e}_{\sigma}^{*}\right\|_{L^2(\Omega)}
\end{aligned}
\tag{54}
$$

During the SPR-CD recovery process, the functional (22) tries to minimize $\left\|\boldsymbol{u}_{u}^{*}-\boldsymbol{u}^{h}\right\|_{L^2(\mathscr{P}^i)}$ at each patch. In addition, as $\boldsymbol{u}_{\sigma}^{*}$ would be a recovered displacement field, laying in a so-called 'broken space' richer than V^h, we could assume that the L^2-norm of the error of the recovered solution is similar to than the L^2 norm of the error of the FE solution:

$$\left\|\boldsymbol{e}_{\sigma}^{*}\right\|_{L^2(\Omega)}\approx\|\boldsymbol{e}\|_{L^2(\Omega)}\tag{55}$$

Considering this assumption in (54) we would obtain an upper error bound of the recovered solution:

$$|||\boldsymbol{e}_{\sigma}^{*}|||_{\Omega}\lesssim\left(\left\|\boldsymbol{s}_{\sigma}^{*}\right\|_{L^2(\Omega)}+\left\|\boldsymbol{r}_{\sigma}^{*}\right\|_{L^2(\Omega)}\right)\|\boldsymbol{e}\|_{L^2(\Omega)}\tag{56}$$

Note that $\|\boldsymbol{e}\|_{L^2(\Omega)}$ is unknown in general. One possibility is to replace it by $\|\boldsymbol{e}_u\|_{L^2(\Omega)}$, recall that $\boldsymbol{e}_u=\boldsymbol{u}_{u}^{*}-\boldsymbol{u}^{h}$, obtaining a computational version of the upper bound. The implicit idea is to replace $\left\|\boldsymbol{e}_{\sigma}^{*}\right\|_{L^2(\Omega)}$ by $\|\boldsymbol{e}_u\|_{L^2(\Omega)}$, i.e. we have replaced the error in the recovered solution $\boldsymbol{u}_\sigma^*$ by the estimated error of the FE solution $\boldsymbol{u}^h$, to obtain a bound of the error in the recovered solution.

Following this idea we can also derive expressions for the error estimator. We can replace $\boldsymbol{e}_\sigma^* = \boldsymbol{u} - \boldsymbol{u}_\sigma^*$ by $\boldsymbol{e}_u = \boldsymbol{u}_u^* - \boldsymbol{u}_h$ in (52), and define the following error estimator $\mathscr{E}_1^*$ in (57a) to check if it could provide an indication of the error level in energy norm of the recovered solution $\boldsymbol{\sigma}_\sigma^*$. We also defined the error indicators $\mathscr{E}_2^*$ and $\mathscr{E}_3^*$ as described in (57b) and (57c) to force positive values into the square root:

$$\mathscr{E}_1^* = \sqrt{-\int_\Omega (\boldsymbol{s}_\sigma^*)^T \boldsymbol{e}_u \, \mathrm{d}\Omega - \int_\Gamma (\boldsymbol{r}_\sigma^*)^T \boldsymbol{e}_u \, \mathrm{d}\Gamma} \tag{57a}$$

$$\mathscr{E}_2^* = \sqrt{\sum_K \left(\left| \int_K (\boldsymbol{s}_\sigma^*)^T \boldsymbol{e}_u \, \mathrm{d}\Omega \right| + \left| \int_{\partial K \cup \Gamma_N} (\boldsymbol{r}_\sigma^*)^T \boldsymbol{e}_u \, \mathrm{d}\Gamma \right| \right)} \tag{57b}$$

$$\mathscr{E}_3^* = \sqrt{\int_\Omega \left| (\boldsymbol{s}_\sigma^*)^T \boldsymbol{e}_u \right| \, \mathrm{d}\Omega + \int_\Gamma \left| (\boldsymbol{r}_\sigma^*)^T \boldsymbol{e}_u \right| \, \mathrm{d}\Gamma} \tag{57c}$$

In (57b) the value of the integrals at each element are forced to be positive. In (57c) the integrands themselves are forced to be positive. Note that this is a reasonable assumption. For example, if we assume $\boldsymbol{r}_\sigma^* = 0$, see (53).

$$0 \leq \boldsymbol{\sigma}(\boldsymbol{e}_\sigma^*)\boldsymbol{\varepsilon}(\boldsymbol{e}_\sigma^*) = -(\boldsymbol{s}_\sigma^*)^T \boldsymbol{e}_\sigma^* = \left| (\boldsymbol{s}_\sigma^*)^T \boldsymbol{e}_\sigma^* \right| \tag{58}$$

As $\boldsymbol{s}_\sigma^*$ and $\boldsymbol{e}_\sigma^*$ are consistent ($\boldsymbol{s}_\sigma^*$ would be the defaults of equilibrium corresponding to $\boldsymbol{u}_\sigma^*$, whose associated error is $\boldsymbol{e}_\sigma^*$) then $0 \leq -(\boldsymbol{s}_\sigma^*)^T \boldsymbol{e}_\sigma^*$. However, in (57c) $\boldsymbol{e}_\sigma^*$ has been substituted by $\boldsymbol{e}_u$. The terms $\boldsymbol{s}_\sigma^*$ and $\boldsymbol{e}_u$ are non-consistent and as a result $-(\boldsymbol{s}_\sigma^*)^T \boldsymbol{e}_u$ could be negative. This suggests the use of the approximation in (57c), $-(\boldsymbol{s}_\sigma^*)^T \boldsymbol{e}_\sigma^* \approx \left| (\boldsymbol{s}_\sigma^*)^T \boldsymbol{e}_u \right|$.

4.2 h-adaptive Refinement Process

In the previous section we have presented several methods to estimate the error of the recovered solution. The numerical results will indicate that the estimator $\mathscr{E}_3^*$ provides very accurate results with an excellent global effectivity index. Now, we are going to show how to use this error estimator of the recovered solution to define a h-adaptive refinement processes.

In the classical situation we estimate the error of the raw FE solution, because use the raw FE solution $(\boldsymbol{u}^h, \boldsymbol{\sigma}^h)$ as output. However, when we use our recovery procedure we have an improved solution $(\boldsymbol{u}_u^*, \boldsymbol{\sigma}_\sigma^*)$ available. So far, we were unable to estimate the error of this last solution, therefore this output was not reliable. However, in the following section we will show that $\mathscr{E}_3^*$ provides a good error indicator of the recovered field. Therefore we can use $(\boldsymbol{u}_u^*, \boldsymbol{\sigma}_\sigma^*)$ as the output of the analysis. The information about the error estimation in the recovered solution could then be used in the h-adaptive refinement process to obtain a solution $(\boldsymbol{u}_u^*, \boldsymbol{\sigma}_\sigma^*)$ with the required accuracy.

The h-adaptive procedure proposed in this section is guided by the well-established techniques based on the error estimation of the FE solution. As a consequence of the refinement process the error of the FE solution and the error of the recovered solution will simultaneously decrease. An scheme of the proposed h-adaptive process was shown in Fig. 4. The main difference between the traditional refinement process and the proposed one is the stopping criterion. We simply propose to stop the h-adaptive refinement process when the estimated error of the recovered solution is smaller than the target error. As the recovered solution reaches the prescribed error level with less degrees of freedom than the FE solution, this method produces important savings in the total computational cost of the analysis. The process would then be as follows:

1. Generate a FE mesh.
2. Solve the FE problem.
3. Evaluate the local error estimate of FE solution $(\boldsymbol{u}^h, \boldsymbol{\sigma}^h)$.
4. Evaluate the global value of the error estimate of the recovered solution $(\boldsymbol{u}_u^*, \boldsymbol{\sigma}_\sigma^*)$ (costless procedure).
5. If target error is smaller than the error of the recovered solution continue to step 6, else stop the process.
6. Generate a h-adapted mesh using the local FE error estimation
7. Go to step 2.

5 Numerical Results

We will illustrate the performance of the techniques described in the previous sections using a numerical example. The problem to be solved corresponds to a thick-wall cylinder under internal pressure. The geometrical model for this problem is represented in Fig. 5. Due to symmetry, only 1/4 of the section is modelled. The internal and external surfaces are of radius a and b. Young's modulus is $E = 1000$, Poisson's ratio is $\nu = 0.3$, $a = 5$, $b = 20$ and the internal pressure $P = 1$. The exact solution

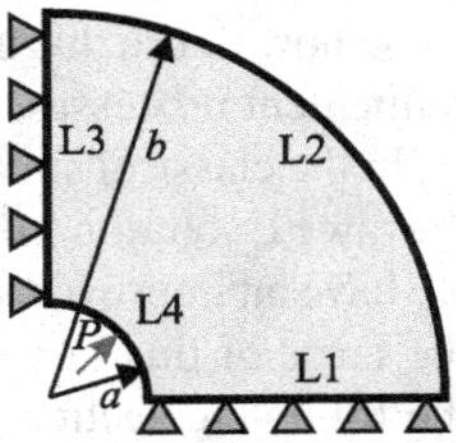

Fig. 5 Problem 2. Thick-wall cylinder under internal pressure

for the radial displacement assuming plane strain conditions is given by:

$$u_r(r) = \frac{P(1+\nu)}{E(c^2-1)}\left(r(1-2\nu) + \frac{b^2}{r}\right) \quad (59)$$

where $c = b/a$, $r = \sqrt{x^2+y^2}$ and $\phi = \arctan(y/x)$. Stresses in cylindrical coordinates are given by:

$$\sigma_r(r) = \frac{P}{c^2-1}\left(1 - \frac{b^2}{r^2}\right) \quad \sigma_\phi(r) = \frac{P}{c^2-1}\left(1 + \frac{b^2}{r^2}\right) \quad \sigma_z(r,\phi) = 2\nu\frac{P}{c^2-1} \quad (60)$$

Figure 6a shows the effectivity θ of the error estimators presented in this work. We observe that $\mathscr{E}_{\text{UB}}$ and $\mathscr{E}_{\text{LB}}$ provide accurate upper and lower error bounds. The error estimator using the SPR-CD technique, $\mathscr{E}_{\text{CD}}$, is also providing very accurate results but bounding properties are not maintained. Regarding to the error indicator of the recovered solution, $\hat{\mathscr{E}}_3^*$, we also observe that accurate results are obtained since the effectivity index is between 0.8 and 1.2. The interest in using the recovered solution as output is illustrated in Fig. 6b where the convergence curves of the exact relative error in energy norm are plotted. We observe that not only an increase of accuracy of more than an order of magnitude is obtained in this problem, but also a higher convergence rate. This fact permits a considerable decrease in the computational cost for a given accuracy, speeding up the resolution of the numerical analysis.

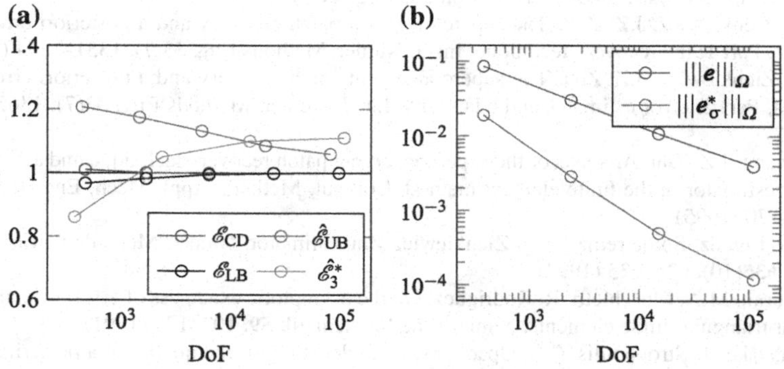

Fig. 6 Uniform refinement with linear elements. Global effectivity index for the different error estimation techniques and convergence of the error in energy norm. **a** Effectivity index θ. **b** Relative error in energy norm

6 Conclusions

We have presented the main aspects of the error estimation techniques based on recovery procedures. The techniques can be easily implemented and provide accurate error estimations but were unable to lead to guaranteed error bounds. However, the development of enhanced recovery techniques, like the SPR-C and SPR-CD techniques has allowed for the development of error bounding techniques based on recovery procedures. The numerical results have shown the accuracy and robustness of the techniques. Since recovery-based error estimation techniques require the evaluation of a recovered field of higher quality that the raw FE solution, the use of the recovered solution as output of the analysis codes instead of the FE solution is of high interest, but requires assessing the accuracy of the recovered solution. We have shown that an error indication technique for the recovered field can be used in h-adaptive refinement processes to reduce the computational cost of the numerical analysis. The promising result show the interest and need of analytical studies in this direction.

References

1. O.C. Zienkiewicz, R.Taylor, *The Finite Element Method*, vol 1, 4th edn. (Oxford, Butterworth-Heinemann, 1989)
2. O.C. Zienkiewicz, J.Z. Zhu, A simple error estimator and adaptive procedure for practical engineering analysis. Int. J. Numer. Methods Eng. **24**(2), 337–357 (1987)
3. O.C. Zienkiewicz, J.Z. Zhu, The superconvergent patch recovery and a posteriori error estimates. Part 1: the recovery technique. Int. J. Numer. Methods Eng. **33**(7), 1331–1364 (1992)
4. O.C. Zienkiewicz, J.Z. Zhu, The superconvergent patch recovery and a posteriori error estimates. Part 2: error estimates and adaptivity. Int. J. Numer. Methods Eng. **33**(7), 1365–1382 (1992)
5. Z. Zhang, J.Z. Zhu, Analysis of the superconvergent patch recovery technique and a posteriori error estimator in the finite element method. Comput. Methods Appl. Mech. Eng. **163**(1–4), 159–170 (1995)
6. R. Rodriguez, Some remarks on Zienkiewicz-Zhu estimator. Numer. Methods Partial Differ. Equ. **635**(10), 625–635 (1994)
7. R. Durán, M.A. Muschietti, R. Rodriguez, On the asymptotic exactness of error estimators for linear triangular finite elements. Numerische Mathematik **59**, 107–127 (1991)
8. I. Babuška, T. Strouboulis, C.S. Upadhyay, A model study of the quality of a posteriori error estimators for linear elliptic problems. Error estimation in the interior of patchwise uniform grids of triangles. Comput. Methods Appl. Mech. Eng. **114**(3–4), 307–378 (1994)
9. I. Babuška, T. Strouboulis, C.S. Upadhyay, S.K. Gangaraj, K. Copps, Validation of a posteriori error estimators by numerical approach. Int. J. Numer. Methods Eng. **37**(7), 1073–1123 (1994)
10. I. Babuška, T. Strouboulis, C.S. Upadhyay, A model study of the quality of a posteriori error estimators for finite element solutions of linear elliptic problems, with particular reference to the behaviour near the boundary. Int. J. Numer. Methods Eng. **40**(14), 2521–2577 (1997)
11. M. Ainsworth, J.Z. Zhu, A.W. Craig, O.C. Zienkiewicz, Analysis of the Zienkiewicz-Zhu a-posteriori error estimator in the finite element method. Int. J. Numer. Methods Eng. **28**(9), 2161–2174 (1989)

12. J.J. Ródenas, M. Tur, F.J. Fuenmayor, A. Vercher, Improvement of the superconvergent patch recovery technique by the use of constraint equations: the SPR-C technique. Int. J. Numer. Methods Eng. **70**(6), 705–727 (2007)
13. J.J. Ródenas, O.A. González-Estrada, P. Díez, F.J. Fuenmayor, Accurate recovery-based upper error bounds for the extended finite element framework. Comput. Methods Appl. Mech. Eng. **199**(37–40), 2607–2621 (2010)
14. E. Nadal, S. Bordas, J.J. Ródenas, J.E. Tarancón, M. Tur, Accurate Stress Recovery for the Two-Dimensional Fixed Grid Finite Element Method, in *Procedings of the Tenth International Conference on Computational Structures Technology*, pp. 1–20 (2010)
15. S.P. Timoshenko, J.N. Goodier, *Theory of Elasticity*, 2nd edn. (McGraw-Hill, New York, 1951)
16. T. Blacker, T. Belytschko, Superconvergent patch recovery with equilibrium and conjoint interpolant enhancements. Int. J. Numer. Methods Eng. **37**(3), 517–536 (1994)
17. O.A. González-Estrada, J.J. Ródenas, S.P.A. Bordas, M. Duflot, P. Kerfriden, E. Giner, On the role of enrichment and statical admissibility of recovered fields in a-posteriori error estimation for enriched finite element methods. Eng. Comput. **29**(8) (2012)
18. J.J. Ródenas, O.A. González-Estrada, J.E. Tarancón, F.J. Fuenmayor, A recovery-type error estimator for the extended finite element method based on singular+smooth stress field splitting. Int. J. Numer. Methods Eng. **76**(4), 545–571 (2008)
19. S.P.A. Bordas, M. Duflot, Derivative recovery and a posteriori error estimate for extended finite elements. Comput. Methods Appl. Mech. Eng. **196**(35–36), 3381–3399 (2007)
20. M. Duflot, S.P.A. Bordas, A posteriori error estimation for extended finite elements by an extended global recovery. Int. J. Numer. Methods Eng. **76**, 1123–1138 (2008)
21. B.A. Szabó, I. Babuška, *Finite Element Analysis* (Wiley, New York, 1991)
22. J.J. Ródenas, E. Giner, J.E. Tarancón, O.A. González-Estrada, A Recovery Error Estimator for Singular Problems Using Singular+Smooth Field Splitting, in *Fifth International Conference on Engineering Computational Technology*, ed. by B.H.V. Topping, G. Montero, R. Montenegro (Civil-Comp Press, Stirling, Scotland, 2006)
23. E. Nadal, Cartesian grid FEM (cgFEM): high performance h-adaptive FE analysis with efficient error control. Application to structural shape optimization. Ph.D. thesis, Universitat Politècnica de València (2014)
24. O.A. González-Estrada, E. Nadal, J.J. Ródenas, P. Kerfriden, S.P.A. Bordas, F.J. Fuenmayor, Mesh adaptivity driven by goal-oriented locally equilibrated superconvergent patch recovery. Comput. Mech. (2013)
25. P.Díez, J.J. Ródenas, O.C. Zienkiewicz, Equilibrated patch recovery error estimates: simple and accurate upper bounds of the error. Int. J. Numer. Methods Eng. **69**(200610), 2075–2098 (2007)
26. P.G. Ciarlet, *The Finite Element Method For Elliptic Problems*, 1st edn. (North-Holland publishing company, Amsterdam, 1978)
27. T. Gerasimov, M. Rüter, E. Stein, An explicit residual-type error estimator for Q 1 -quadrilateral extended finite element method in two-dimensional linear elastic fracture mechanics. Int. J. Numer. Methods Eng. **90**, 1118–1155 (2012)
28. C. Carstensen, S.A. Funken, Constants in Clément-Interpolation error and residual-based a posteriori estimates in finite element methods. East-West J. Numer. Math. **8**(3), 153–175 (2000)
29. P. Díez, N. Parés, A. Huerta, Recovering lower bounds of the error by postprocessing implicit residual a posteriori error estimates. Int. J. Numer. Methods Eng. **56**(10), 1465–1488 (2003)
30. N.E. Wiberg, F. Abdulwahab, Error estimation with postprocessed finite element solutions. Comput. Struct. **64**(1–4), 113–137 (1997)
31. I. Babushka, T. Strouboulisb, S.K. Gangarajb, A posteriori estimation of the error in the recovered derivatives the finite element solution. Comput. Method **150**, 369–396 (1997)

The Constitutive Relation Error Method: A General Verification Tool

Pierre Ladevèze and Ludovic Chamoin

Abstract This chapter reviews the Constitutive Relation Error method as a general verification tool which is very suitable to compute strict and effective error bounds for linear and more generally convex Structural Mechanics problems. The review is focused on the basic features of the method and the most recent developments.

Keywords A posteriori error estimation · Constitutive relation error · Duality · Goal-oriented control · Nonlinear problems · PGD models

1 Introduction

Today, more than ever, modeling and simulation are central to any mechanical engineering activity. A constant concern both in industry and in research has always been the verification of models which nowadays can attain very high levels of complexity. The novelty of the situation is that over the last thirty years truly quantitative tools for assessing the quality of a FE model have appeared; this topic is now known as *model verification*. Of course, the original continuum mechanics model remains the reference. The state of the art can be found in [1–5]. Until the late 90s, only global error estimators were available through three different families introduced by [6–8]. Besides error indicators, adaptive computational approaches related to the mesh, time and iteration parameters have been developed for nearly all problems in Structural Mechanics. The CRE-verification method could be seen as a unified method in this context.

Since 1990, a key issue has become the evaluation of the quality of outputs of interest resulting from a finite element analysis. This objective was beyond the reach

P. Ladevèze (✉) · L. Chamoin
LMT (ENS Cachan, CNRS, Paris-Saclay University),
61 Avenue du Président Wilson, 94230 Cachan, France
e-mail: ladeveze@lmt.ens-cachan.fr

L. Chamoin
e-mail: chamoin@lmt.ens-cachan.fr

L. Chamoin and P. Díez (eds.), *Verifying Calculations – Forty Years On*,
SpringerBriefs in Applied Sciences and Technology,
DOI 10.1007/978-3-319-20553-3_4

of earlier error estimators, which provided only global information which was totally insufficient for mechanical design purposes (design criteria involve local values of stresses, displacements, stress intensity factors,...). Among the numerous works on the linear case, we can mention [9–13] as the earliest ones; further references can be found in [2–5]. The main idea which emerged then was that an output of interest can be written globally, thus allowing the reuse of global error estimators; however, accurate error estimation requires the finite element solution of what is called the *adjoint problem*. Extensions to nonlinear time-dependent cases appeared in the late 90s [12, 14, 15]; these approaches consisted in getting back to the linear case through linearization during each time step.

Unfortunately, most of these estimates are not guaranteed to be upper or lower bounds, which is a very serious drawback for robust design. Consequently, one of today's research challenges is to derive upper error bounds for the calculated values of outputs of interest. Even in the linear case, relatively few works have proposed answers [1, 4, 9–11, 13, 16, 17]. Outside of the FE context, and only for the linear case, there are a few early works on this subject, such as [18, 19]. These, which use analytical Green functions, have serious limitations and seem quite remote from the present concern, in which numerical aspects are central.

Recent papers [20–22] introduced new upper error bounds on a computed output of interest for linear as well as time-dependent nonlinear problems, even in dynamics. These were probably the first strict upper bounds published for nonlinear and transient dynamics cases. Small-displacement problems without softening, such as (visco-)plasticity, were included through the standard thermodynamics framework involving internal state variables. Classical convexity properties were assumed. These works completed the a posteriori error estimation method developed particularly at LMT-Cachan, which was based on the concept of Constitutive Relation Error (CRE) of the dissipation type and on quasi-explicit techniques for the construction of associated admissible FE solutions.

The first key point of this approach was the integration of an output of interest in terms of finite variations; this led to the introduction of what is called the *mirror problem at time T*, which is very similar to the initial problem, as a substitute for the adjoint problem. Of course, the mirror problem coincides with the adjoint problem in the linear case. Another key point concerned the convexity properties, which constitute the true "engine" of our approach for deriving upper error bounds. These properties led to the basic relations between the dissipation-type constitutive relation error and the solution error. Eventually, upper error bounds were derived on the basis of the data and the FE solutions of both reference and mirror problems over the time-space domain being studied. It is also important to note that these bounds take all sources of error into account: time and space discretizations, ending of the iterations, and also modeling errors.

The present chapter goes in two directions. First, we give the main features of the CRE-method, going into details only for the linear case. Second, we introduce the advances performed recently. They mainly concern key technical points [23–27] and also various engineering problems [21, 22, 28–33]. Complex material models even in dynamics are considered and Constitutive Relation Errors are developed in [4].

The idea is rather simple; all equations are satisfied by admissible fields except the Constitutive Relation, so that the value of the residue related to the verification of the constitutive relation is an error indicator of the quality of the approximate solution. In other words, the approximate solution could be seen as the exact solution of the problem with a modified constitutive relation; then we compare this modified constitutive relation with the reference one. Here, we focus on upper error bounds and therefore only a particular class of material models is investigated, class characterized by convexity properties.

2 Reference Problem and Notations

Initially, the structure being studied occupies a domain $\Omega \subset \mathbb{R}^d$ with boundary $\partial\Omega$ (Fig. 1). We assume small displacements, quasi-static loading and isothermal conditions. The time interval of interest is denoted $[0, T]$. At any time t belonging to $[0, T]$, the structure is placed in an environment characterized by a displacement $\mathbf{U}_d$ on a part $\partial_1\Omega \subset \partial\Omega$, traction forces $\mathbf{F}_d$ on $\partial_2\Omega$ (the part of $\partial\Omega$ complementary to $\partial_1\Omega$), and body forces $\mathbf{f}_d$ within the domain Ω.

The problem which describes the evolution of the structure over $[0, T]$ is:

Find the displacement field $\mathbf{u}(\mathbf{x}, t)$ and the stress field $\sigma(\mathbf{x}, t)$, with $t \in [0, T]$ and $\mathbf{x} \in \Omega$, which verify:

- the kinematic constraints:

$$\mathbf{u} \in \mathscr{U}^{[0,T]} \; ; \quad \mathbf{u}_{|\partial_1\Omega} = \mathbf{U}_d \text{ on }]0, T[\tag{1}$$

- the equilibrium equations (principle of virtual work):

$$\sigma \in \mathscr{S}^{[0,T]} \; ; \forall t \in]0, T[\;\; \forall \mathbf{u}^* \in \mathscr{U}_{ad,0}$$

$$-\int_\Omega \sigma : \varepsilon(\mathbf{u}^*) \mathrm{d}\Omega + \int_\Omega \mathbf{f}_d \cdot \mathbf{u}^* \mathrm{d}\Omega + \int_{\partial_2\Omega} \mathbf{F}_d \cdot \mathbf{u}^* \mathrm{d}S = \int_\Omega \rho \frac{\mathrm{d}^2\mathbf{u}}{\mathrm{d}t^2} \cdot \mathbf{u}^* \mathrm{d}\Omega \tag{2}$$

- the constitutive relation:

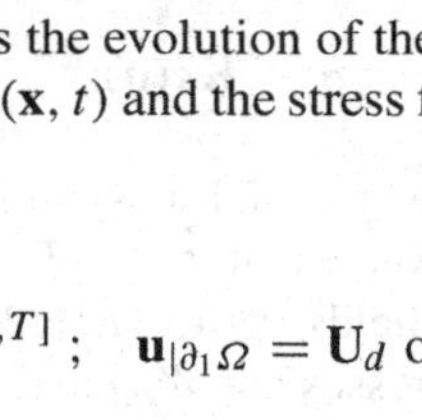

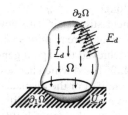

Fig. 1 Schematic representation of the environment (i.e. the prescribed conditions)

$$\forall t \in [0,T] \;\; \forall \mathbf{x} \in \Omega \quad \sigma_{|t} = \mathbf{A}\big(\varepsilon(\dot{\mathbf{u}}_{|\tau});\, \tau \leq t\big) \tag{3}$$

$\varepsilon(\mathbf{u})$ denotes the strain associated with the displacement $\big(\varepsilon(\mathbf{u})_{ij} = \frac{1}{2}(u_{i,j} + u_{j,i})\big)$. $\mathcal{U}^{[0,T]}$ is the space containing the displacement field $\mathbf{u}$ defined over $\Omega \times]0,T[$, and $\mathcal{S}^{[0,T]}$ is the space containing the stresses, also defined over $\Omega \times]0,T[$. Finally, $\mathcal{U}_{ad,0}$ is the vector space of the prescribed virtual velocities. Operator $\mathbf{A}$, which is given and generally single-valued, characterizes the mechanical behavior of the material. ρ is the density.

In the following we denote $\mathcal{U}_{ad}^{[0,T]}$ the space of displacement fields which verify (1), and $\mathcal{S}_{ad}^{[0,T]}$ the space of stress fields which verify (2).

3 Use of CRE for Elasticity Problems

Let us start with the simplest family of mechanical problems, i.e. elasticity problems. We focus on the final state of the structure at $t = T$; thus, the problem to be solved does not depend on time. Moreover, the constitutive relation (3) becomes:

$$\sigma = \mathbf{K}\varepsilon(\mathbf{u}) \tag{4}$$

where $\mathbf{K}$ denotes the Hooke tensor, which is symmetric and positive definite. The densities $\mathbf{U}_d$, $\mathbf{f}_d$, and $\mathbf{F}_d$ are known at $t = T$. $\mathcal{U} = [H^1(\Omega)]^d$ and $\mathcal{S} = \{$second order symmetric tensor fields $\pi \in [L^2(\Omega)]^{d(d+1)/2}\}$. The spaces of the admissible displacements and stresses are $\mathcal{U}_{ad}$ and $\mathcal{S}_{ad}$, respectively.

3.1 Basics on CRE and Global Error Estimation

We assume that the finite element solution was calculated using a displacement approach. Thus, the finite element displacement-stress pair $(\mathbf{u}_h, \sigma_h)$ is known and the stress σ_h is FE-equilibrated.

The principle behind our approach consists in associating a new and admissible displacement-stress pair (i.e. belonging to $\mathcal{U}_{ad} \times \mathcal{S}_{ad}$), denoted $(\hat{\mathbf{u}}_h, \hat{\sigma}_h)$, to the data and the finite element displacement-stress pair. This new entity also verifies what we call the *prolongation conditions*, which are relations with the finite element solution. The construction of $(\hat{\mathbf{u}}_h, \hat{\sigma}_h)$ is achieved through a general quasi-explicit technique which is now well-known [4, 34, 35], and which has several recent variants [26, 36]; an overview is given in the Appendix. For the displacement field, we generally take $\hat{\mathbf{u}}_h = \mathbf{u}_h$; for quasi-incompressible materials, a modification is shown in [37].

Let us first recall the Prager-Synge theorem [4, 38] which links the constitutive relation error to the error in the solution. Introducing the constitutive relation error:

$$[E^h_{CRE}]^2 \equiv \Phi^*(\hat{\sigma}_h) + \Phi(\hat{\varepsilon}_h) - \int_\Omega \hat{\sigma}_h : \hat{\varepsilon}_h \mathrm{d}\Omega \quad \text{with} \quad \hat{\varepsilon}_h \equiv \varepsilon(\hat{\mathbf{u}}_h) \tag{5}$$

where Φ and Φ^* are the global potentials (dual convex functions) of the constitutive relation:

$$\Phi^*(\sigma) \equiv \int_\Omega \varphi^*(\sigma)\mathrm{d}\Omega \quad ; \quad \varphi^*(\sigma) \equiv \frac{1}{2}\sigma : \mathbf{K}^{-1}\sigma$$
$$\Phi(\varepsilon) \equiv \int_\Omega \varphi(\varepsilon)\mathrm{d}\Omega \quad ; \quad \varphi(\varepsilon) \equiv \frac{1}{2}\varepsilon : \mathbf{K}\varepsilon \tag{6}$$

we have the following properties:

$$\Phi^*(\sigma - \hat{\sigma}_h) + \Phi(\varepsilon - \hat{\varepsilon}_h) = [E^h_{CRE}]^2$$
$$\Phi^*(\sigma - \hat{\sigma}_{h,m}) = \frac{1}{4}[E^h_{CRE}]^2 \quad \text{with} \quad \hat{\sigma}_{h,m} \equiv \frac{1}{2}[\hat{\sigma}_h + \mathbf{K}\hat{\varepsilon}_h] \tag{7}$$

The first relation in (7) leads to the guaranteed upper bound $||\mathbf{u} - \mathbf{u}_h||_\mathbf{K} \leq \sqrt{2}E^h_{CRE}(\mathbf{u}_h, \hat{\sigma}_h)$ on the global discretization error (in the energy norm $|| \bullet ||_\mathbf{K}$), and the quality of this bound depends on that of $\hat{\sigma}_h$. An illustration taken from [36] is given in Fig. 2 and exhibits the distribution of the error estimate $\sqrt{2}E^h_{CRE}(\mathbf{u}_h, \hat{\sigma}_h)$ over the elements of the mesh. The structure is clamped on part of its boundary and subjected to unit traction forces on the opposite boundary. The elastic material is isotropic and linear with $E = 1$ and $\nu = 0.3$.

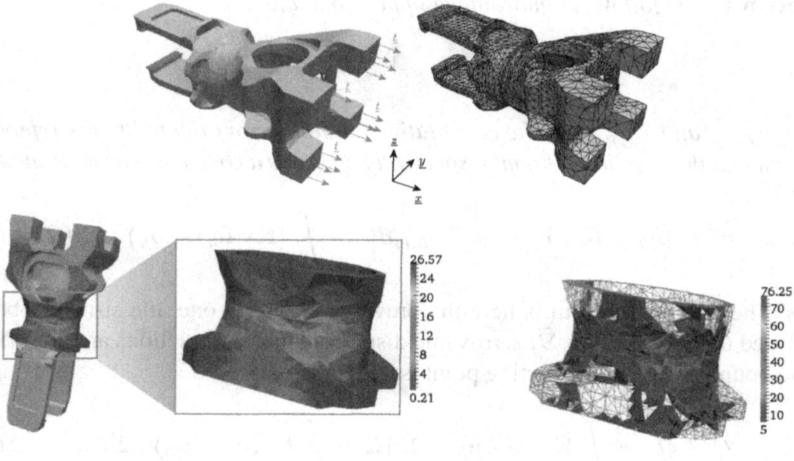

Fig. 2 Reference problem (*top left*) with associated FE mesh (*top right*), magnitude of the FE Von Mises stress (*bottom left*), and distribution of higher local contributions to the error (*bottom right*)

3.2 Goal-Oriented Error Estimation

We now consider a local measure of the discretization error, defined on an output of interest Q.

3.2.1 The Output of Interest

The quantity of interest Q is a goal-oriented quantity, such as the mean value of a stress or displacement component over an element or a set of elements. Such an output of interest can be written globally:

$$Q = \int_\Omega \sigma : \tilde{\Sigma} \mathrm{d}\Omega \tag{8}$$

where $\tilde{\Sigma}$ is the extractor which defines Q. Here, for the sake of simplicity, we do not consider convex nonlinear functionals of the stress, but such extensions would not involve serious difficulties. The Q-error is $|Q - Q_h|$ where Q_h is the value obtained from the finite element solution $(\mathbf{u}_h, \sigma_h)$.

In order to get a relevant bound on $|Q - Q_h|$, a common practice is to calculate what is called the adjoint problem, which is very classical in this case: it consists in an elasticity problem with a pre-strain $\tilde{\Sigma}$. Its finite element solution, which can be obtained with a refined mesh, is denoted $(\tilde{\mathbf{u}}_h, \tilde{\sigma}_h)$, and an associated admissible FE solution is $(\hat{\tilde{\mathbf{u}}}_h, \hat{\tilde{\sigma}}_h)$. The main result is [20]:

Theorem 1 *The following guaranteed upper bound holds:*

$$|Q - Q_h - Q_{corr}| \leq E^h_{CRE}.\tilde{E}^h_{CRE} \tag{9}$$

where E^h_{CRE} and $\tilde{E}^h_{CRE}$ are the constitutive relation errors related to the reference problem and the adjoint problem, respectively. Q_{corr} is a correction term defined as:

$$Q_{corr} \equiv \int_\Omega \big(\hat{\sigma}_h - \mathbf{K}\varepsilon(\hat{\mathbf{u}}_h)\big) : \mathbf{K}^{-1}\hat{\tilde{\sigma}}_{h,m} d\Omega + \int_\Omega \big(\mathbf{K}\varepsilon(\hat{\mathbf{u}}_h) - \sigma_h\big) : \tilde{\Sigma} d\Omega \tag{10}$$

Proof The adjoint problem is here the previous elasticity one, the structure being submitted to the pre-strain $\tilde{\Sigma}$; zero-value displacements and tractions are prescribed on the boundary $\partial\Omega$. The starting point is:

$$Q - Q_h = \int_\Omega \mathbf{K}\varepsilon(\mathbf{u} - \hat{\mathbf{u}}_h) : \tilde{\Sigma} \mathrm{d}\Omega + \int_\Omega \mathbf{K}\varepsilon(\hat{\mathbf{u}}_h - \mathbf{u}_h) : \tilde{\Sigma} \mathrm{d}\Omega \tag{11}$$

Noticing that:

$$\begin{aligned}\int_\Omega \mathbf{K}\varepsilon(\mathbf{u}-\hat{\mathbf{u}}_h) : \tilde{\Sigma}\mathrm{d}\Omega &= \int_\Omega \varepsilon(\mathbf{u}-\hat{\mathbf{u}}_h) : \mathbf{K}\varepsilon(\tilde{\mathbf{u}})\mathrm{d}\Omega \\ &= \int_\Omega \varepsilon(\mathbf{u}-\hat{\mathbf{u}}_h) : (\tilde{\sigma} - \mathbf{K}\varepsilon(\hat{\tilde{\mathbf{u}}}_h))\mathrm{d}\Omega + \int_\Omega \varepsilon(\mathbf{u}-\hat{\mathbf{u}}_h) : \mathbf{K}\varepsilon(\hat{\tilde{\mathbf{u}}}_h)\mathrm{d}\Omega \quad (12) \\ &= \int_\Omega \varepsilon(\mathbf{u}-\hat{\mathbf{u}}_h) : (\hat{\tilde{\sigma}}_h - \mathbf{K}\varepsilon(\hat{\tilde{\mathbf{u}}}_h))\mathrm{d}\Omega + \int_\Omega (\hat{\sigma}_h - \mathbf{K}\varepsilon(\hat{\mathbf{u}}_h)) : \varepsilon(\hat{\tilde{\mathbf{u}}}_h)\mathrm{d}\Omega\end{aligned}$$

and introducing $\hat{\sigma}_{h,m}$ and $\hat{\tilde{\sigma}}_{h,m}$ as in (7), we get $Q - Q_h - Q_{corr} = \int_\Omega \left(\sigma - \hat{\sigma}_{h,m}\right) : \mathbf{K}^{-1}\left(\hat{\tilde{\sigma}}_h - \mathbf{K}\varepsilon(\hat{\tilde{\mathbf{u}}}_h)\right)\mathrm{d}\Omega$. Using the Cauchy-Schwarz inequality and (7), we obtain the final upper bound.

Remark 1 The bound defined by the second member of (9) is half of the classical bound (see [4]). Indeed, introducing $\hat{\sigma}_{h,m}$ enables a more accurate bounding.

Remark 2 The value of the calculated error bound depends on the meshes used to solve the reference and adjoint problems. It is always possible, by refining the mesh of the adjoint problem alone, to control the value of the Q-error. In general, a local refinement (near the domain of interest) of the mesh used to solve the adjoint problem is very effective [23, 33].

Remark 3 The Galerkin orthogonality property related to the FE solution is not used to derive Theorem 1.

Remark 4 Theorem 1 is based on the Cauchy-Schwarz inequality as nearly all error bounds. In [27], a new bounding technique using the Saint-Venant principle and homothetic domains is derived; this can give sharper bounds. The idea is to decompose the domain Ω in two disjoint zones: (i) zone ω_λ, parameterized with scalar value λ, surrounding the zone where the quantity of interest is defined; (ii) complementary zone Ω/ω_λ. We can then write $Q - Q_h - Q_{corr} = q_{\omega_\lambda} + q_{\Omega/\omega_\lambda}$.

Bounding the term $q_{\Omega/\omega_\lambda}$ can be easily and accurately performed from the Cauchy-Schwarz inequality applied over Ω/ω_λ, as $\tilde{E}^h_{CRE|\Omega/\omega_\lambda}$ remains small in practice. Bounding the other term q_{ω_λ} is more technical; it leans on an inequality, related to Saint-Venant's principle, of the form:

$$||\sigma - \hat{\sigma}_h||_{\mathbf{K}^{-1}|\omega_\lambda} \leq \left(\frac{\lambda}{\overline{\lambda}}\right)^{1/k} ||\sigma - \hat{\sigma}_h||_{\mathbf{K}^{-1}|\omega_{\overline{\lambda}}} + \gamma_{\lambda,\overline{\lambda}} \quad (13)$$

where $\omega_{\overline{\lambda}}$ is a homothetic domain of ω_λ, parameterized by scalar value $\overline{\lambda} \geq \lambda$ (Fig. 3), k is a computable constant that depends on the geometry of ω_λ (and obtained analytically or numerically by solving an additional local eigenvalue problem), and $\gamma_{\lambda,\overline{\lambda}}$ is a known term. In practice, $\overline{\lambda}$ is chosen the highest possible while ensuring $\omega_{\overline{\lambda}} \subset \Omega$, and λ the smallest possible with ω_λ surrounding the zone of interest. The exponential decrease with respect to $\lambda/\overline{\lambda}$ in (13) is the key point to avoid overestimation.

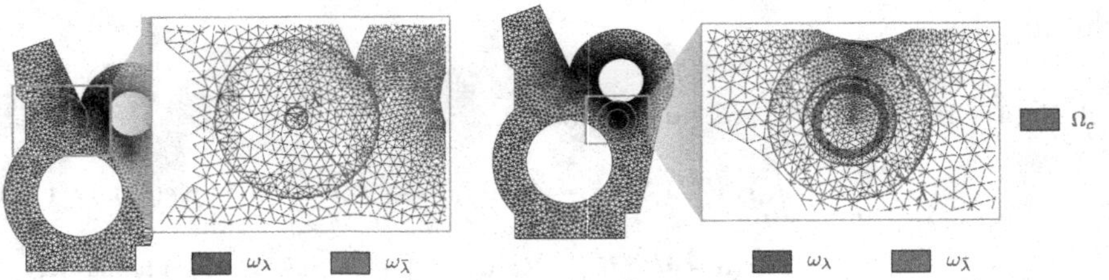

Fig. 3 Homothetic domains ω_λ and $\omega_{\overline{\lambda}}$ defined in a cracked structure when considering different quantities of interest: local mean of a component of σ (*left*), and stress intensity factors in the vicinity of the crack (*right*)

4 Use of CRE for Visco-Plasticity Problems

4.1 Basics on CRE and Global Error Estimation

We rewrite the reference problem (1)–(3) using some global notations within the framework of classical thermodynamics with internal variables. Let us introduce the following generalized quantities:

$$s = \begin{bmatrix} \sigma \\ \mathbf{Y} \end{bmatrix} \quad \dot{e}_p = \begin{bmatrix} \dot{\varepsilon}_p \\ -\dot{\mathbf{X}} \end{bmatrix} \quad \dot{e}_e = \begin{bmatrix} \dot{\varepsilon}_e \\ \dot{\mathbf{X}} \end{bmatrix} \tag{14}$$

where additional internal variables are gathered in the n-vectors $\mathbf{X}$ and $\mathbf{Y}$. The dissipation bilinear form over the time-space domain is:

$$(\dot{e}_p, s) \longmapsto \int_0^T \int_\Omega s \cdot \dot{e}_p \,\mathrm{d}\Omega\,\mathrm{d}t = \int_0^T \int_\Omega \big(\sigma : \dot{\varepsilon}_p - \mathbf{Y} \cdot \dot{\mathbf{X}}\big)\mathrm{d}\Omega\,\mathrm{d}t \tag{15}$$

The reference problem is thus to find $(\dot{e}_p, s) \in \mathscr{S}^{[0,T]}$ such that:

$$\begin{aligned}
&\bullet\ \dot{e} = \dot{e}_e + \dot{e}_p \text{ is kinematically admissible}\\
&\bullet\ s \text{ is statically admissible}\\
&\bullet\ e_e = \boldsymbol{\Lambda}(s) \quad \text{(state equations)}\\
&\bullet\ \dot{e}_p = \mathbf{B}(s) \quad \text{(evolution laws)}\\
&\bullet\ s = 0, e = 0 \text{ at } t = 0 \quad \text{(initial conditions)}
\end{aligned} \tag{16}$$

$\boldsymbol{\Lambda}$ is assumed to be linear; most viscoplastic materials are in this category (see [3, 39]) and this material description is called *normal*. $\mathbf{B}$ could be nonlinear and multivalued, as in plasticity, but here we consider the family of standard materials whose state evolution laws can be expressed with two potentials, which are dual (in the Legendre-Fenchel sense) convex functions φ and φ^* such that for $(t, \mathbf{x}) \in [0, T] \times \Omega$:

$$\forall(\dot{e}_p, s) \in \mathscr{S}^{[0,T]} \quad \varphi^*(s) + \varphi(\dot{e}_p) - s \cdot \dot{e}_p \geq 0$$
$$\varphi^*(s) + \varphi(\dot{e}_p) - s \cdot \dot{e}_p = 0 \Longleftrightarrow \dot{e}_p = \mathbf{B}(s) \qquad (17)$$

Then, the constitutive relation (dissipation) error related to $(\dot{e}_p, s) \in \mathscr{S}^{[0,T]}$ is:

$$[E_{CRE}]^2 = \int_0^T \int_\Omega \Big[\varphi^*(s) + \varphi(\dot{e}_p) - s \cdot \dot{e}_p\Big] \mathrm{d}\Omega \, \mathrm{d}t \qquad (18)$$

4.1.1 The Associated Admissible FE Solution

In the dissipation error framework, the concept of admissibility must be modified [3]:

Definition 1 A pair $(\dot{e}_p, s) \in \mathscr{S}^{[0,T]}$ is admissible if:
(i) the state equations are verified $\big(e_e = \boldsymbol{\Lambda}(s)\big)$
(ii) $\dot{e} = \dot{e}_e + \dot{e}_p$ and s verify the kinematic constraints and the equilibrium equations.

To go further, let us assume that the calculated solution was obtained using the FE method. Thus, at discrete time points t_m belonging to $[0, T]$, we know:

$$[\dot{e}_h, s_h]_t \quad ; \quad t \in [0, t_1, \dots, t_n = T] \qquad (19)$$

and $[\dot{e}_h, s_h]_t$ verifies the kinematic constraints and equilibrium equations in the FE sense at these discrete time points. Assuming that the evolution of the data during each time step is linear, we can extend the FE solution across the whole time-space domain. Thus, we get $(\dot{e}_h, s_h) \in \mathscr{S}^{[0,T]}$ which verifies the kinematic constraints and the equilibrium equations in the FE sense at any time $t \in [0, T]$.

In order to get an associated admissible solution, we use the same technique as in elasticity to define a displacement-stress pair $(\hat{\mathbf{u}}_h, \hat{\sigma}_h)$ which is admissible in the classical sense i.e. which verifies the kinematic constraints, the equilibrium equations and the initial conditions over $[0, T] \times \Omega$. Let us note that in the case of (visco-)plasticity with the constraint $\mathrm{Tr}[\dot{\varepsilon}_p] = 0$, the previous displacement must be modified so that $\mathrm{Tr}[\dot{\hat{\varepsilon}}_p] = 0$. The additional internal variables $(\hat{\mathbf{X}}_h, \hat{\mathbf{Y}}_h)$ which must verify the state equations can be easily constructed by solving local problems related to the minimization of the constitutive relation error of the dissipation type. Finally, we obtain an admissible solution $(\dot{\hat{e}}_p, \hat{s}) \in \mathscr{S}^{[0,T]}$ of the reference model. More details can be found in [4, 39].

The differences between the computerized structural model and the reference model are not limited to numerical aspects like time and space discretizations; models can differ in their state evolution laws. For example, the reference can be a viscoplastic material model while the calculations are performed with an elastic model.

4.1.2 Link Between the Constitutive Relation Error and the Error in the Solution

Let us first introduce what we call the φ-tangent potential at x:

$$\overline{\varphi}(x' - x) \equiv \varphi(x') - \varphi(x) - y \cdot (x' - x) \geq 0 \tag{20}$$

where φ, φ^* are two dual convex functions and (x, y) is such that $\varphi(x) + \varphi^*(y) - x \cdot y = 0$; $\overline{\varphi} = \varphi$ for quadratic potentials. We give a new version of the fundamental link between the constitutive relation error and the error in the solution derived in [3]:

Theorem 2 *Let $(\dot{e}_p, s)$ be the exact solution and $(\dot{\hat{e}}_{p,h}, \hat{s}_h)$ an arbitrary admissible solution to the reference problem. We have:*

$$\overline{\Phi}^*\big(s - \mathbf{B}^{-1}(\dot{\hat{e}}_{p,h})\big) + \overline{\Phi}\big(\dot{e}_p - \mathbf{B}(\hat{s}_h)\big) + \int_0^T |\dot{a}| E_F(s - \hat{s}_h)dt = [E^h_{CRE}]^2 \tag{21}$$

$$\text{with : } \bullet\ \overline{\Phi} \equiv \int_0^T \int_\Omega a(t)\overline{\varphi}\, d\Omega\, dt \qquad \overline{\Phi}^* \equiv \int_0^T \int_\Omega a(t)\overline{\varphi}^* d\Omega\, dt$$

$$\bullet\ [E^h_{CRE}]^2 = \int_0^T \int_\Omega a(t)\Big[\varphi^*(\hat{s}_h) + \varphi(\dot{\hat{e}}_{p,h}) - \hat{s}_h \cdot \dot{\hat{e}}_{p,h}\Big] d\Omega\, dt \tag{22}$$

- *E_F : free energy*
- *$a(t)$: arbitrary function such that $a(t) \geq 0$ $\dot{a} \leq 0$ $a(T) = 0$*

In some applications, it is interesting to restrict the time interval to a subinterval $[T', T]$. An identity similar to (21) holds, with the additional term $a(T')E^+_{F|T'}$ on the right-hand side, where E^+_F is an upper bound on $E_F(s - \hat{s}_h)_{|T'}$.

Proposition 1 *An upper bound on the free energy $E_F(s - \hat{s}_h)$ at t is $E^+_F(t)$, solution to:*

$$E^+_F(0) = 0\ ;\quad \frac{d}{dt}(E^+_F) + \omega(E^+_F, t) = \int_\Omega \Big[\varphi^*(\hat{s}_h) + \varphi(\dot{\hat{e}}_{p,h}) - \hat{s}_h \cdot \dot{\hat{e}}_{p,h}\Big] d\Omega \tag{23}$$

where ω is a function such that:

$$\omega(E^+_F(\Delta s), t) \leq \inf_{\Delta s \in \mathscr{S}_{ad,0}} \int_\Omega \Big[\overline{\varphi}^*\big(\Delta s + \hat{s}_h - \mathbf{B}^{-1}(\dot{\hat{e}}_{p,h})\big) + \overline{\varphi}\big(\mathbf{B}(\Delta s + \hat{s}_h) - \mathbf{B}(\hat{s}_h)\big)\Big] d\Omega \tag{24}$$

$\mathscr{S}_{ad,0}$ being the space of statically admissible generalized stress fields under homogeneous conditions.

The proof of Proposition 1 essentially uses Theorem 2, written locally in time.

4.2 Goal-Oriented Error Estimation

The output of interest can be defined as:

$$Q = \int_0^T \int_\Omega [\sigma : \delta\dot{\Sigma} - \mathbf{Y} \cdot \delta\dot{\mathbf{X}}] \mathrm{d}\Omega \, \mathrm{d}t = \int_0^T \int_\Omega s \cdot \delta\dot{\tilde{e}}_\Sigma \, \mathrm{d}\Omega \, \mathrm{d}t \tag{25}$$

where the extractor is $\delta\dot{\tilde{e}}_\Sigma = \begin{bmatrix} \delta\dot{\Sigma} \\ -\delta\dot{\mathbf{X}} \end{bmatrix}$ with $\delta\dot{\tilde{e}}_\Sigma = 0$ at $t = T$.

Here, δ is a symbol indicating that $\delta\dot{\tilde{e}}_\Sigma$ must be interpreted as a finite, but relatively small, variation. *However, we do not carry out any linearization.* The classical adjoint problem is replaced by what we call the *mirror* problem, which is similar to the reference problem except that time goes backwards: $\tau \equiv T - t$. This mirror problem, written with δ-quantities, is defined as:

Find $(\delta\dot{\tilde{e}}_p, \delta\tilde{s}) \in \mathscr{S}^{[0,T]}$ such that:

$$\begin{aligned}
&\bullet\ \delta\dot{\tilde{e}} = \delta\dot{\tilde{e}}_e + \delta\dot{\tilde{e}}_p \text{ is kinematically admissible} \\
&\bullet\ \delta\tilde{s} - \delta\tilde{s}_\Sigma \text{ is statically admissible} \\
&\bullet\ \delta\tilde{e}_e = \boldsymbol{\Lambda}(\delta\tilde{s}) \quad \text{(state equations)} \\
&\bullet\ \delta\dot{\tilde{e}}_p = \tilde{\mathbf{B}}(\delta\tilde{s}) \equiv \boldsymbol{B}(s_h + \delta\tilde{s}) - \mathbf{B}(s_h) \quad \text{(evolution laws)} \\
&\bullet\ \delta\tilde{s} = 0, \delta\dot{\tilde{e}} = 0 \text{ at } \tau = 0 \quad \text{(initial conditions)}
\end{aligned} \tag{26}$$

where $s_h(\tau)$ is the FE solution to the reference problem and $\delta\dot{\tilde{e}}_\Sigma = \tilde{\mathbf{B}}(\delta\tilde{s}_\Sigma) + \boldsymbol{\Lambda}(\delta\dot{\tilde{s}}_\Sigma)$.

Let $(\dot{\tilde{e}}_h, \tilde{s}_h)$ be the FE solution to the mirror problem and $(\dot{\hat{\tilde{e}}}_h, \hat{\tilde{s}}_h)$ the associated admissible FE solution over $[0, T] \times \Omega$. From now on, all these quantities will be defined with respect to the initial time t.

4.2.1 Upper Error Bound for an Output of Interest

The starting point is the following relation, which can be easily proven.

Proposition 2 *Using the previous notations, the Q-error is equal to:*

$$-Q + Q_h + Q_{corr} = \int_0^T \int_\Omega (s - \hat{s}_{h,m}) \cdot (\tilde{\mathbf{B}}(\delta\hat{\tilde{s}}_h) - \delta\dot{\hat{\tilde{e}}}_{p,h}) d\Omega \, dt + \mathbf{C}(s - s_h, \delta\tilde{s}_\Sigma) - \mathbf{C}(s - s_h, \delta\hat{\tilde{s}}_h) \tag{27}$$

$$where: \bullet\ Q_{corr} \equiv -\int_0^T \int_\Omega \Big[(\hat{\dot{e}}_h - \dot{e}_h)\cdot(\delta\tilde{s}_\Sigma - \delta\hat{\tilde{s}}_h) - (\hat{s}_h - s_h)\cdot\delta\dot{\hat{\tilde{e}}}_h$$
$$+ (\hat{s}_{h,m} - s_h)\cdot(\tilde{\mathbf{B}}(\delta\hat{\tilde{s}}_h) - \delta\dot{\hat{\tilde{e}}}_{p,h})\Big]d\Omega\, dt$$
$$\bullet\ \dot{e}_h = \boldsymbol{A}(\dot{s}_h) + \mathbf{B}(s_h) \quad ; \quad \delta\dot{\hat{\tilde{e}}}_{p,h|t} = \big[\delta\tilde{e}_{h,\tau} - \boldsymbol{A}(\delta\hat{s}_{h,\tau})\big]_{|\tau=T-t}$$
$$\bullet\ \mathbf{C}(\Delta s, \delta s) = \int_0^T \int_\Omega [\Delta s\cdot\delta\dot{e}_p - \Delta\dot{e}_p\cdot\delta s]d\Omega\, dt \quad with \quad \Delta\dot{e}_p = \tilde{\mathbf{B}}(\Delta s) \tag{28}$$

Remark 5 The idea behind this relation is related to the **C**-terms: these are very small if the finite variations $(s - s_h)$ and $\delta\tilde{s}_\Sigma$ are small. Moreover, if the material model is linear, the **C**-terms are equal to zero. **C** is called the *model nonlinearity indicator*.

Remark 6 The generalized stress $\hat{s}_{h,m}$ is similar to the mean stress introduced in (7). In practice, we take an approximation of the minimization problem related to the cost function g:

$$g(s - \hat{s}_{h,m}) = \int_0^T \Big[a\,\overline{\varphi}^*\big(s - \mathbf{B}^{-1}(\dot{\hat{e}}_{p,h})\big) + a\,\overline{\varphi}\big(\mathbf{B}(s) - \mathbf{B}(\hat{s}_h)\big) + |\dot{a}|E_F(s - \hat{s}_h)\Big]dt \tag{29}$$

Remark 7 The Galerkin orthogonality conditions related to the FE solution are not used.

We now derive upper bounds on the two terms in the right-hand side of (27). We only give the results here, but more details can be found in [20, 29].

Proposition 3 *Let be* $I_1 = \int_0^T \int_\Omega (s - \hat{s}_{h,m})\cdot(\tilde{\mathbf{B}}(\delta\hat{\tilde{s}}_h) - \delta\dot{\hat{\tilde{e}}}_{p,h})d\Omega\, dt$. *We have:*

$$I_1 \leq 2\Big[[E^h_{CRE}]^2 - [E^h_{CRE,m}]^2\Big]^{\frac{1}{2}} \cdot \Big[F^2(\overline{\mu}\,\dot{\tilde{a}}_h)\Big]^{\frac{1}{2}} + F^1(\dot{\tilde{a}}_h) \tag{30}$$

$$with: \bullet\ [E^h_{CRE,m}]^2 = \int_\Omega [\min_{y\in\mathscr{F}^{[0,T]}} g(y)]d\Omega \qquad \dot{\tilde{a}}_h = \tilde{\mathbf{B}}(\delta\hat{\tilde{s}}_h) - \delta\dot{\hat{\tilde{e}}}_h$$
$$\bullet\ f(\dot{x}) \equiv \sup_{y\in\mathscr{F}^{[0,T]}} \Big[\int_0^T y\cdot\dot{x}\, dt - g(y)\Big] \quad \forall\dot{x}\in\mathscr{E}^{[0,T]} \quad \textit{(Legendre-Fenchel transform of g)}$$
$$\bullet\ f(\mu\,\dot{x}) = f(0) + \mu f^1(\dot{x}) + f^2(\mu\,\dot{x}) \quad with\ \mu \geq 0 \quad \lim_{\mu\to 0^+}\frac{f^2(\mu\,\dot{x})}{\mu} = 0 \tag{31}$$
$$\bullet\ F(\cdot) = \int_\Omega f(\cdot)d\Omega \qquad 1 = \frac{\big[[E^h_{CRE}]^2 - [E^h_{CRE,m}]^2\big]}{F^2(\overline{\mu}\,\dot{\tilde{a}}_h)}$$

An alternative bound for I_1 consists in using the Legendre-Fenchel transform for the dissipation bilinear form written at $(t, \mathbf{x}) \in [0, T] \times \Omega$. This bound is easier to obtain, but it is less effective. Another option is to work globally over $[0, T] \times \Omega$.

Let us now consider $I_2 \equiv \mathbf{C}(s - s_h, \delta\tilde{s}_\Sigma) - \mathbf{C}(s - s_h, \delta\hat{s}_h)$. We introduce $g_\Delta(\Delta s) \equiv g(\Delta s + \hat{s}_{h,m} - s_h)$ and define analytically, or at least numerically:

$$f_\Delta(\dot{a}, -b) = \sup_{\substack{\Delta s \in \mathscr{F}^{[0,T]} \\ \Delta\dot{e}_p = \tilde{\mathbf{B}}(\Delta s)}} \int_0^T \big[\Delta s \cdot \dot{a} - \Delta\dot{e}_p \cdot b - g_\Delta(\Delta s)\big] \mathrm{d}t \qquad (32)$$

Writing $F_\Delta(\mu\delta\dot{\tilde{a}}, -\mu\delta\tilde{b}) = F_\Delta(0,0) + \mu F^1_\Delta(\delta\dot{\tilde{a}}, -\delta\tilde{b}) + F^2_\Delta(\mu\delta\dot{\tilde{a}}, -\mu\delta\tilde{b})$, the following bound can be proved:

$$I_2 \leq \big[[E^h_{CRE}]^2 - [E^h_{CRE,m}]^2\big]^{\frac{1}{2}} \big[F^2_\Delta(\overline{\mu}\delta\dot{\tilde{a}}, -\overline{\mu}\delta\tilde{b})\big]^{\frac{1}{2}} + F^1_\Delta(\delta\dot{\tilde{a}}, -\delta\tilde{b}) \quad ; \quad 1 = \frac{\big[[E^h_{CRE}]^2 - [E^h_{CRE,m}]^2\big]}{F^2_\Delta(\overline{\mu}\delta\dot{\tilde{a}}, -\overline{\mu}\delta\tilde{b})} \qquad (33)$$

F_Δ is small if the finite variations Δs and δs are small. A similar technique can be derived to get a strict lower bound.

Finally, from previous results, we obtain strict upper bounds on $|Q - Q_h - Q_{corr}|$. Illustrations of such bounds can be found in the literature: viscoelasticity (taking history effects into account) was addressed in [23], dynamics problems were considered in [22, 31], and nonlinear viscoplasticity problems were studied in [21]. Furthermore, the general case where the material operator $\mathbf{B}$ is not defined using two dual potentials (convex functions) but is simply maximum monotonous is given in [29].

In the following section, we focus on some recent advances for the computation of both accurate and practical bounds, in addition to be guaranteed, using the CRE framework.

5 Getting Accurate and Practical Error Bounds on Outputs of Interest

5.1 Non-intrusive Approach for the Adjoint Solution

As noticed in Sect. 3.2, the accuracy of the error bounds on Q can be controlled solving the adjoint problem with a locally refined mesh. However, this has the drawback to require remeshing of the structure. An alternative, qualified as non-intrusive as the initial mesh is not changed, was proposed in [24, 25]. It consists in a local enrichment of the adjoint solution, using PUM, with known *handbook* functions that aim at representing the high gradient part of $(\tilde{\mathbf{u}}, \tilde{\sigma})$; the approach is thus similar to that proposed in XFEM or GFEM [40, 41] except that no additional dof is introduced here (the singularity comes from the adjoint loading). We present the method in the linear elasticity case, but extensions to time-dependent problems can be found in [22, 24].

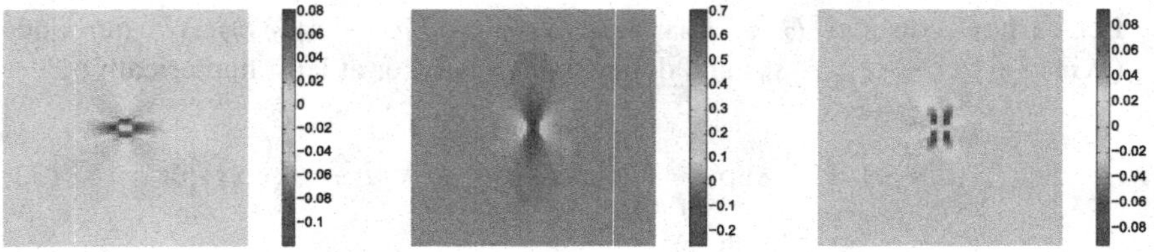

Fig. 4 Quasi-exact stress field in a (quasi-) infinite domain subjected to a local prestress in a squared region: $\tilde{\sigma}_{xx}^{hand}$ *(left)*, $\tilde{\sigma}_{yy}^{hand}$ *(center)*, $\tilde{\sigma}_{xy}^{hand}$ *(right)*

Enrichment functions which are used, denoted $(\tilde{\mathbf{u}}^{hand}, \tilde{\sigma}^{hand})$ in the following, correspond to generalized Green's functions and represent the (quasi-)exact adjoint solution in a (semi-)infinite domain subjected to the adjoint loading. These functions are obtained either analytically (when possible) or pre-computed numerically with a fine mesh an sufficiently large domain. An example of such a function, corresponding to a localized prestress loading $\tilde{\sigma}_\Sigma$, is given on Fig. 4. *Handbook* functions are inserted in the adjoint solution using the partition of unity defined by linear FE shape functions N_i associated to vertices i of the initial mesh $(N_i(\mathbf{x}_j) = \delta_{ij})$. The enrichment is introduced locally, and consequently only a set of n^{PUM} vertices are used. The enrichment region $\Omega^{PUM} \subset \Omega$ is defined as $\{\mathbf{x} \in \Omega, \sum_{i=1}^{n^{PUM}} N_i(\mathbf{x}) \neq 0\}$; it can be divided in two disjoint subregions Ω_1^{PUM} and Ω_2^{PUM} such that:

$$\sum_{i=1}^{n^{PUM}} N_i(\mathbf{x}) = \begin{cases} 1 & \text{in } \Omega_1^{PUM} \\ a \in]0,1[& \text{in } \Omega_2^{PUM} \\ 0 & \text{in } \Omega/\overline{(\Omega_1^{PUM} \cup \Omega_2^{PUM})} \end{cases} \tag{34}$$

In practice, Ω_1^{PUM} is such that it contains the zone of interest Ω_Σ in which quantity Q is defined, i.e. the region that supports extraction functions.

Therefore, the displacement solution to the adjoint problem is searched under the form:

$$\tilde{\mathbf{u}}(\mathbf{x}) = \sum_{i=1}^{n^{PUM}} \tilde{\mathbf{u}}^{hand}(\mathbf{x}) N_i(\mathbf{x}) + \tilde{\mathbf{u}}^{res}(\mathbf{x}) \tag{35}$$

where $\tilde{\mathbf{u}}^{res}$ is a residual solution, usually very regular, to be computed. The enrichment part $\sum_{i=1}^{n^{PUM}} \tilde{\mathbf{u}}^{hand} N_i$ enables to reproduce local high gradients of $\tilde{\mathbf{u}}$ whereas the residual part $\tilde{\mathbf{u}}^{res}$ enables to correct the enrichment part in order to verify boundary conditions of the adjoint problem on $\partial\Omega$. The new expression of $\tilde{\sigma}$ is deducted from (35):

$$\tilde{\sigma}(\mathbf{x}) = \tilde{\sigma}_{PUM}^{hand}(\mathbf{x}) + \tilde{\sigma}^{res}(\mathbf{x}) \tag{36}$$

with $\tilde{\sigma}_{PUM}^{hand} = \mathbf{K}\varepsilon(\sum_{i=1}^{n^{PUM}} \tilde{\mathbf{u}}^{hand} N_i)$ and $\tilde{\sigma}^{res} = \mathbf{K}\varepsilon(\tilde{\mathbf{u}}^{res})$; of course, $\tilde{\sigma}_{PUM}^{hand} = \tilde{\sigma}^{hand}$ in Ω_1^{PUM}.

Once the set of n^{PUM} enriched vertices are defined, the new adjoint problem consists in finding $\tilde{\mathbf{u}}^{res} \in \mathcal{U}_{ad,0}$ such that:

$$a(\mathbf{v}, \tilde{\mathbf{u}}^{res}) = Q(\mathbf{v}) - a(\mathbf{v}, \sum_{i=1}^{n^{PUM}} \tilde{\mathbf{u}}^{hand} N_i) \quad \forall \mathbf{v} \in \mathcal{U}_{ad,0} \tag{37}$$

The residual stress field $\tilde{\sigma}^{res} = \mathbf{K}\varepsilon(\tilde{\mathbf{u}}^{res})$ then verifies the following balance equation:

$$\begin{aligned}\int_{\Omega} \tilde{\sigma}^{res} : \varepsilon(\mathbf{v})\mathrm{d}\Omega &= \int_{\Omega} \left(\tilde{\sigma}_{\Sigma} : \varepsilon(\mathbf{v}) + \tilde{\mathbf{f}}_{\Sigma} \cdot \mathbf{v}\right) \mathrm{d}\Omega - \int_{\Omega} \tilde{\sigma}_{PUM}^{hand} : \varepsilon(\mathbf{v})\mathrm{d}\Omega \\ &= \int_{\Omega_1^{PUM}} \left(\tilde{\sigma}_{\Sigma} : \varepsilon(\mathbf{v}) + \tilde{\mathbf{f}}_{\Sigma} \cdot \mathbf{v} - \tilde{\sigma}^{hand} : \varepsilon(\mathbf{v})\right) \mathrm{d}\Omega - \int_{\Omega_2^{PUM}} \tilde{\sigma}_{PUM}^{hand} : \varepsilon(\mathbf{v})\mathrm{d}\Omega \\ &= - \int_{\partial\Omega_1^{PUM}} \tilde{\sigma}^{hand} \mathbf{n}_{12} \cdot \mathbf{v}\mathrm{d}S - \int_{\Omega_2^{PUM}} \tilde{\sigma}_{PUM}^{hand} : \varepsilon(\mathbf{v})\mathrm{d}\Omega \quad \forall \mathbf{v} \in \mathcal{U}_{ad,0}\end{aligned} \tag{38}$$

where $\mathbf{n}_{12}$ is the outgoing normal from Ω_1^{PUM} to Ω_2^{PUM}. The loading consists in tractions $-\tilde{\sigma}^{hand}\mathbf{n}_{12}$ on $\partial\Omega_1^{PUM}$ and a prestress $-\tilde{\sigma}_{PUM}^{hand}$ in Ω_2^{PUM}.

A fine approximation $(\tilde{\mathbf{u}}_h^{res}, \tilde{\sigma}_h^{res})$ of the residual solution can be obtained with the initial mesh; the enrichment technique is thus non-intrusive in the sense where operators (stiffness matrix, mesh connectivities) defined for the primal problem can be reused without any change to solve the adjoint problem; only the loading has to be modified. The computation of an admissible stress field $\hat{\tilde{\sigma}}_h$ is also performed in a non-intrusive way: one first defines a stress field $\hat{\tilde{\sigma}}_h^{res}$ that verifies (38), with the same method as that used to compute $\hat{\sigma}_h$; we then get:

$$\hat{\tilde{\sigma}}_h(\mathbf{x}) = \tilde{\sigma}_{PUM}^{hand}(\mathbf{x}) + \hat{\tilde{\sigma}}_h^{res}(\mathbf{x}) \tag{39}$$

Eventually, we obtain the following bounding:

$$|Q - Q_h - Q_{corr}| \leq E_{CRE}^h . \tilde{E}_{CRE,res}^h \tag{40}$$

where the right-hand side is independent of the enrichment $(\tilde{\mathbf{u}}^{hand}, \tilde{\sigma}^{hand})$. In practice, the error on the residual solution $\tilde{E}_{CRE,res}^h$ is small, and (40) provides for very accurate bounds on the local error without remeshing.

5.2 Application to Pointwise Quantities

When considering a quantity of interest which is pointwise in space (and/or in time), the loading of the adjoint problem is defined from Dirac functions. Consequently, the corresponding adjoint solution is highly singular and possibly infinite energy;

consequently, it is useless to compute an approximate solution with the FEM. A classical alternative method consists in regularizing the quantity of interest, using for instance weighting functions (molifiers) [11], in order to preserve the regularity of the adjoint solution; the initial quantity of interest is then replaced with a weighted local average.

The non-intrusive enrichment technique presented previously enables, with direct extension, to consider the evaluation of the discretization error for truly pointwise quantities, without resorting to regularization. For that, Green functions have to be chosen as enrichment functions. Under some assumptions, these functions can be determined analytically in a (semi-)infinite domain [42, 43]; an example of such a function is given in Fig. 5. In more complex situations (anisotropic material for instance), Green functions can sometimes be obtained implicitly [44].

Bounding (40) is still valid for pointwise quantities Q and bounds can be calculated despite of the fact that $(\tilde{\mathbf{u}}^{hand}, \tilde{\sigma}^{hand})$ may be infinite energy; indeed, only the residual part $(\tilde{\mathbf{u}}_h^{res}, \hat{\tilde{\sigma}}_h^{res})$ of the admissible adjoint solution is used to obtain error bounds. Nevertheless, technical numerical tools (numerical integration) are required to compute Q_{corr} and the loading term of the adjoint problem that imply the singular *handbook* solution $(\tilde{\mathbf{u}}^{hand}, \tilde{\sigma}^{hand})$.

In [22], the non-intrusive approach was conducted for visco-elastodynamics problems. Dealing with the singularity of the adjoint solution in space and time was performed by means of a local enrichment with Green's function associated to Q. Such a function can be calculated in a (semi-)infinite medium from the correspondence principle and Laplace transform. It was shown that local error bounds deteriorate when the model tends to a pure elastodynamics model; in this case, the influence zone of the Green function occupies the whole domain Ω and one needs to represent correctly this function after reflection on the boundary $\partial\Omega$; when viscous effects are important, this zone remains localized as the magnitude rapidly tends to zero (see Fig. 6 where we observe wave fronts P and S without or with 20 % damping, with Dirac loading in space and time). Local error bounds also deteriorate when the quantity of interest becomes more and more localized in time; this illustrates that considering a pointwise quantity in time for dynamics problems does not make real sense.

Eventually, and in addition to the non-intrusive approach, a technique was introduced in [25] in order to conserve guaranteed error bounds on nonlinear pointwise

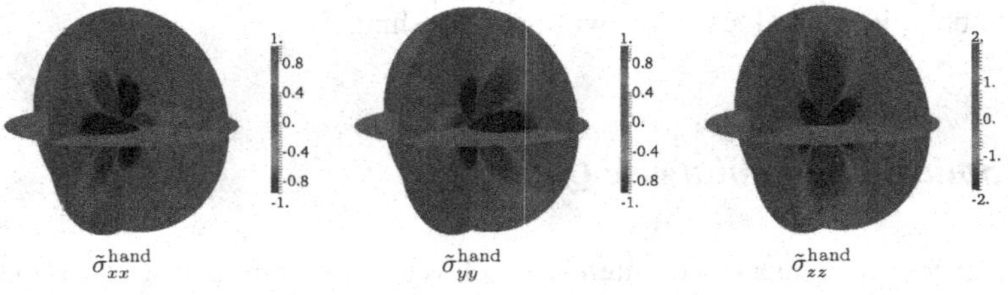

Fig. 5 Stress field associated to a pointwise prestress in a 3D infinite domain

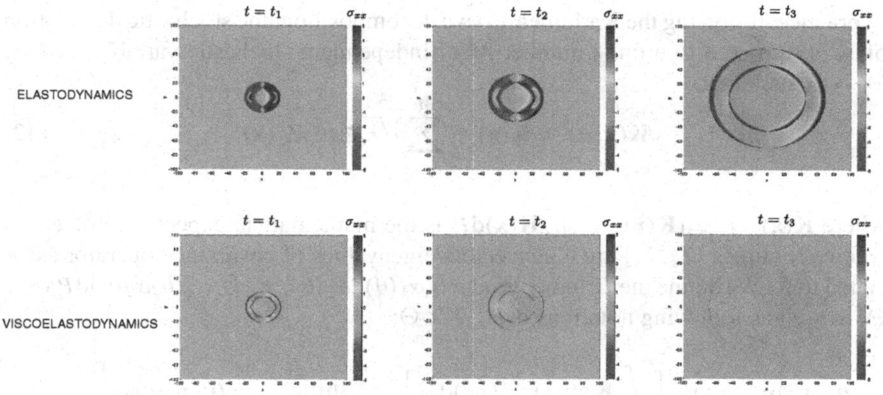

Fig. 6 Evolution of the xx component of *handbook* stress field for a model without (*top*) or with (*bottom*) damping

quantities of interest (such as the Von Mises equivalent stress). This technique is based on a decomposition of Q with projection properties in order to take higher order terms into account in the bounds. It leads to the introduction of a set of extractors, and therefore to the solution to a set of adjoint problems.

6 Control of Various Computational Approaches

6.1 Stochastic Approach

Currently, numerous models involving stochastic parameters are used to take hazards or lack of knowledge into account in numerical simulations. The verification of such models using the CRE concept has been investigated in [28, 45]; it requires: (i) the construction of admissible fields in a stochastic framework; (ii) the splitting between two error sources, coming from discretizations in the physical and stochastic spaces, in order to drive an adaptive process effectively.

Considering a linear elastic material with fluctuating properties, the Hooke tensor is modeled as a stochastic field $\mathbf{K}(\mathbf{x}, \theta) \in [L^2(\Theta, C^0(\Omega))]^{d^4}$; $(\Theta, \mathscr{F}, P)$ is the Kolmogorov probability space, with Θ the set of possible outcomes, $\mathscr{F}$ a σ-algebra of events (subspaces of Θ), and $P : \mathscr{F} \rightarrow [0, 1]$ a probability measure. We assume that field $\mathbf{K}(\mathbf{x}, \theta)$ is bounded and uniformly coercive, i.e. $\exists(K_{min}, K_{max}) \in]0, +\infty[^2$ such that:

$$0 < K_{min} \leq |\mathbf{K}(\mathbf{x}, \theta)| \leq K_{max} \quad \forall \mathbf{x} \in \Omega, \text{ almost surely} \tag{41}$$

In practice, following the Karhunen-Loeve decomposition, the stochastic description of $\mathbf{K}$ is restricted to a finite number M of independent stochastic variables $\xi_i(\theta)$: $\Theta \rightarrow \mathbb{R}$ such that:

$$\mathbf{K}(\mathbf{x}, \theta) \approx \overline{\mathbf{K}}(\mathbf{x}) + \sum_{i=1}^{M} \sqrt{\lambda_i}\xi_i(\theta)\mathbf{Z}_i(\mathbf{x}) \tag{42}$$

where $\overline{\mathbf{K}}(\mathbf{x}) = \mathrm{E}_\Theta[\mathbf{K}(\mathbf{x})] = \int_\Theta \mathbf{K}(\mathbf{x})\mathrm{d}P$ is the mathematical expectation of $\mathbf{K}(\mathbf{x})$, whereas couples $\{\mathbf{Z}_i, \lambda_i\}$ are eigenvectors/eigenvalues of covariance operator associated to $\mathbf{K}$. We define the L^2 inner product $\langle\alpha_1(\theta), \alpha_2(\theta)\rangle \equiv \int_\Theta \alpha_1(\theta)\alpha_2(\theta)\mathrm{d}P$ over Θ, as well as following notations over $\Omega \times \Theta$:

$$\langle \mathbf{u}_1, \mathbf{u}_2\rangle_{\mathbf{K},\Theta} = \mathrm{E}_\Theta\Big[\int_\Omega \mathbf{K}\varepsilon(\mathbf{u}_1) : \varepsilon(\mathbf{u}_2)\mathrm{d}\Omega\Big] \quad ; \quad ||\mathbf{u}||^2_{\mathbf{K},\Theta} = \langle \mathbf{u}, \mathbf{u}\rangle_{\mathbf{K},\Theta}$$
$$\langle \sigma_1, \sigma_2\rangle_{\mathbf{K}^{-1},\Theta} = \mathrm{E}_\Theta\Big[\int_\Omega \mathbf{K}^{-1}\sigma_1 : \sigma_2\mathrm{d}\Omega\Big] \quad ; \quad ||\sigma||^2_{\mathbf{K}^{-1},\Theta} = \langle \sigma, \sigma\rangle_{\mathbf{K}^{-1},\Theta} \tag{43}$$

Kinematic and static admissibility conditions thus respectively read:

$$\mathbf{u} \in \mathscr{U} \quad ; \quad \mathbf{u}_{|\partial_1\Omega} = \mathbf{u}_d \quad \text{almost surely} \tag{44}$$

$$\sigma \in \mathscr{S} \quad ; \quad \mathrm{E}_\Theta\left[\int_\Omega \sigma : \varepsilon(\mathbf{v})\mathrm{d}\Omega - \int_\Omega \mathbf{f}_d \cdot \mathbf{v}\mathrm{d}\Omega - \int_{\partial_2\Omega} \mathbf{F}_d \cdot \mathbf{v}\mathrm{d}S\right] = 0 \quad \forall \mathbf{v} \in \mathscr{U}_{ad,0} \tag{45}$$

with $\mathscr{U} = [L^2(\Theta, H^1(\Omega))]^d$ and $\mathscr{S} = \left\{\pi; \pi = \pi^T, \pi \in [L^2(\Theta, L^2(\Omega))]^{d^2}\right\}$. An approximate solution $(\mathbf{u}_{h,s}, \sigma_{h,s})$, with $\sigma_{h,s} = \mathbf{K}\varepsilon(\mathbf{u}_{h,s})$, is computed using FEM; without lost of generality, we consider here a non-intrusive technique on the stochastic domain, based on interpolation (discretization parameter s) of a set of computed realizations:

$$\mathbf{u}_{h,s}(\mathbf{x}, \theta) = \sum_k \mathbf{u}^k_{h,s}(\mathbf{x}).\Psi_k(\theta) \quad ; \quad \sigma_{h,s}(\mathbf{x}, \theta) = \sum_k \sigma^k_{h,s}(\mathbf{x}).\Psi_k(\theta) \tag{46}$$

where Ψ_k is a polynomial basis associated to the set $\{\xi_i(\theta)\}_{i=1}^M$ of stochastic variables, defined as $\Psi_k = \prod_{i=1}^M H_{k_i}(\xi_i)$ with $H_{k_i}(\xi_i)$ some orthogonal polynomials.

The definition of CRE in the stochastic framework:

$$E_{CRE}(\hat{\mathbf{u}}, \hat{\sigma}) = ||\hat{\sigma} - \mathbf{K}\varepsilon(\hat{\mathbf{u}})||_{\mathbf{K}^{-1},\Theta} \tag{47}$$

associated with an admissible solution in the sense of (44) and (45) enables to naturally extend properties of Sect. 3.1. An admissible solution $(\hat{\mathbf{u}}_{h,s}, \hat{\sigma}_{h,s})$ can be recovered by a post-processing of $(\mathbf{u}_{h,s}, \sigma_{h,s})$. In particular, from $\sigma^k_{h,s}(\mathbf{x})$, we can construct an associated equilibrated field $\hat{\sigma}^k_{h,s}(\mathbf{x})$ using the same techniques as those defined in the deterministic case [26, 36].

For goal-oriented error estimation, defining the output of interest in a global form:

$$Q = \mathrm{E}_{\Theta}\left[\int_{\Omega} \sigma : \tilde{\Sigma}\,\mathrm{d}\Omega\right] \tag{48}$$

we compute an approximate solution $(\tilde{\mathbf{u}}_{h,s}, \tilde{\sigma}_{h,s})$ of the associated (stochastic) adjoint problem, then an admissible solution. We thus get the following bound:

$$|Q - Q_{h,s} - Q_{corr}| \leq E_{CRE}^{h,s} . \tilde{E}_{CRE}^{h,s} \tag{49}$$

the proof of (49) being similar to the deterministic case. Another bounding consists in applying the Cauchy-Schwarz inequality over the space domain alone, before integrating over the stochastic dimension; the obtained bound is sharper but is more complex to compute.

The discretization error $Q - Q_{h,s}$ comes from two sources: (i) space discretization using a FE mesh; (ii) discretization of the stochastic domain. We can estimate the contribution of each source; indeed the local error can be recast as:

$$Q - Q_{h,s} = [Q - Q_h] + [Q_h - Q_{h,s}] = \Delta Q_{spa} + \Delta Q_{sto} \tag{50}$$

where ΔQ_{spa} (resp. ΔQ_{sto}) is the contribution to the local error due to discretization in the physical (resp. stochastic) space.

On the one hand, ΔQ_{sto} can be estimated with the CRE framework considering a semi-discretized reference model, already discretized in space (exact solution $\mathbf{u}_h$). In the sense of this new model, an admissible solution denoted $(\hat{\mathbf{u}}_s, \hat{\sigma}_s)$ is constructed from a post-processing of $(\mathbf{u}_{h,s}, \sigma_{h,s})$; in particular, $\hat{\sigma}_s$ should verify the FE equilibrium over Θ. A similar construction is used to construct an admissible solution $(\hat{\tilde{\mathbf{u}}}_s, \hat{\tilde{\sigma}}_s)$ for the adjoint problem. Consequently, ΔQ_{sto} can be assessed from the bound $E_{CRE}^{s} \cdot \tilde{E}_{CRE}^{s}$. On the other hand, $\Delta Q_{spa} \approx Q_s - Q_{h,s}$ can be assessed considering a semi-discretized reference problem, already discretized over the stochastic dimension (exact solution $\mathbf{u}_s$). An admissible solution denoted $(\hat{\mathbf{u}}_h, \hat{\sigma}_h)$ is constructed in the sense of this new model; in particular, $\hat{\sigma}_h$ should verify equilibrium for each computed realization in Θ. After computing a similar admissible solution $(\hat{\tilde{\mathbf{u}}}_h, \hat{\tilde{\sigma}}_h)$ for the adjoint problem, ΔQ_{spa} can be assessed from the bound $E_{CRE}^{h} \cdot \tilde{E}_{CRE}^{h}$.

6.1.1 XFEM Approach

We consider here the XFEM method in a two-dimensional setting. Introduced in [40] as a generalization of FEM, XFEM enables to capture local solution features in a cracked domain (Fig. 7). In order to improve the convergence rate and use a non-conforming mesh with respect to a crack Γ (supposed free of charge), the XFEM method consists in enriching the classical FE approximation by means of the partition

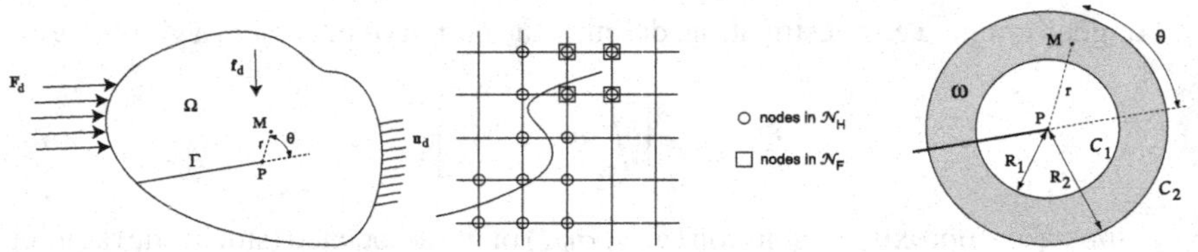

Fig. 7 2D cracked structure (*left*), sets $\mathcal{N}_H$ and $\mathcal{N}_F$ of enriched nodes in a regular mesh (*center*), and definition of the crown ω surrounding the crack tip (*right*)

of unity defined by shape functions $N_i(\mathbf{x})$, and implying two kinds of enrichment functions:

- the Heaviside function $H(\mathbf{x})$ that enables to represent displacement jumps across Γ. This function is ± 1 on the two edges of the crack and is in practice obtained with level-set functions. The enrichment with $H(\mathbf{x})$ is applied to the set $\mathcal{N}_H$ of nodes associated to elements which are cut by the crack (Fig. 7);
- the set $\{F_j(\mathbf{x}), 1 \leq j \leq 4\}$ of basis functions that generate the singular asymptotic solution in the vicinity of the crack tip P:

$$\left\{\sqrt{r}\cos(\frac{\theta}{2}), \sqrt{r}\sin(\frac{\theta}{2}), \sqrt{r}\cos(\frac{\theta}{2})\sin(\frac{\theta}{2}), \sqrt{r}\sin(\frac{\theta}{2})\sin(\frac{\theta}{2})\right\} \tag{51}$$

Enrichment with functions $F_j(\mathbf{x})$ is applied to the set $\mathcal{N}_F$ of nodes associated to elements surrounding the crack tip (Fig. 7).

Therefore, noticing $\mathcal{N}$ the set of nodes in the mesh, the approximate XFEM displacement solution is searched under the form:

$$\mathbf{u}_h(\mathbf{x}) = \sum_{i \in \mathcal{N}} N_i(\mathbf{x})\mathbf{u}_i + \sum_{i \in \mathcal{N}_F} N_i(\mathbf{x}) \left(F_j(\mathbf{x})\mathbf{a}_i^j\right) + \sum_{i \in \mathcal{N}_H} N_i(\mathbf{x})H(\mathbf{x})\mathbf{b}_i \tag{52}$$

where $\{\mathbf{u}_i\}$ is the set of standard FE dofs whereas $\{\mathbf{a}_i^j, \mathbf{b}_i\}$ is the set of additional dofs.

In most cases, compared to classical FEM, the enrichment introduced by XFEM improves the accuracy of the approximate solution as well as values of related quantities of interest. Nevertheless, estimating the discretization error remains fundamental for robust computations. For fracture Mechanics, verification with respect to outputs of interest was addressed in [46, 47] in the context of classical FEM. In the XFEM framework, a guaranteed procedure for local error estimation was introduced and analyzed in [33]. Based on CRE, the technical point of this procedure is again the construction of admissible fields from the XFEM formulation, and it is easy to show that this construction can be carried out by generalizing the classical procedure detailed in the Appendix.

We choose as quantities of interest the stress intensity factors K_I and K_{II}, and use asymptotic extraction functions proposed in [48] and defined over a crown ω surrounding the crack tip:

$$K_\alpha = \int_\omega \varepsilon(\mathbf{u}) : \big(\mathbf{K}\varepsilon(\phi \mathbf{v}_\alpha) - \phi\sigma_\alpha\big)\mathrm{d}\omega - \int_\omega \mathbf{u}\cdot(\sigma_\alpha \nabla\phi)\mathrm{d}\omega \qquad \alpha = I, II \tag{53}$$

In (53), $\mathbf{v}_\alpha$ and σ_α are singular solutions of the reference problem whereas ϕ is a continuous function, defined in the crown ω, that vanishes along the internal boundary $\mathscr{C}_1$ and is 1 along the external boundary $\mathscr{C}_2$. In the following, we consider a circular crown and we use a linear function ϕ defined in polar coordinates as:

$$\begin{cases} \phi(r,\theta) = \frac{R_2 - r}{R_2 - R_1} & \text{for } R_1 \leq r \leq R_2 \\ \quad\quad\;\; = 0 & \text{otherwise} \end{cases} \tag{54}$$

R_1 and R_2 are radii of circles $\mathscr{C}_1$ and $\mathscr{C}_2$, respectively (Fig. 7).

The error estimate (9) remains valid provided admissible solutions $(\hat{\mathbf{u}}_h, \hat{\sigma}_h)$ and $(\hat{\tilde{\mathbf{u}}}_h, \hat{\tilde{\sigma}}_h)$ are constructed from XFEM approximate solutions. On the one hand, the admissible kinematic field $\hat{\mathbf{u}}_h$ (resp. $\hat{\tilde{\mathbf{u}}}_h$) is chosen equal to the approximate displacement $\mathbf{u}_h$ (resp. $\tilde{\mathbf{u}}_h$). On the other hand, the construction of $\hat{\sigma}_h$ (or $\hat{\tilde{\sigma}}_h$) is performed dividing the structure Ω into two complementary zones Ω_1 and Ω_2 in order to take the two kinds of enrichment used in XFEM into account separately. A similar approach was investigated in [32] with a mesh conforming to the crack.

Zone Ω_2, that surrounds the crack tip, contains the set of nodes which are enriched with functions $F_j(\mathbf{x})$. In this zone, $\hat{\sigma}_h$ is built from Airy functions, i.e. expressing components of $\hat{\sigma}_h$ in the polar basis as:

$$\begin{cases} \hat{\sigma}_{h,rr} = \frac{1}{r^2}\phi_{,\theta\theta} + \frac{1}{r}\phi_{,r} \\ \hat{\sigma}_{h,\theta\theta} = \phi_{,rr} \\ \hat{\sigma}_{h,r\theta} = -(\frac{1}{r}\phi_{,\theta})_{,r} \end{cases} ; \quad \phi(r,\theta) = \sum_{i=1}^{n} r^{\beta_i+2}\gamma_i(\theta) \tag{55}$$

with $\beta_i = \frac{1}{2}(i-2)$. Functions γ_i $(i = 1,\ldots,n)$, that correspond to asymptotic solutions, read:

$$\gamma_i(\theta) = A_i \sin(\beta_i\theta) + B_i\cos(\beta_i\theta) + C_i\sin((\beta_i+2)\theta) + D_i\cos((\beta_i+2)\theta) \tag{56}$$

and constants A_i, B_i, C_i and D_i verify:

$$\begin{cases} B_i + D_i = 0 \text{ and } A_i i + C_i(i+2) = 0 & \text{for } \beta_i = 0,1,2,\ldots \\ A_i + C_i = 0 \text{ and } B_i(i+1/2) + D_i(i+5/2) = 0 & \text{for } \beta_i = -1/2, 1/2, \ldots \end{cases} \tag{57}$$

Optimal values of constants A_i, B_i, C_i and D_i are then determined by solving a minimization problem over Ω_2:

$$\{A_i, B_i, C_i, D_i\} = argmin_{\{A'_i,B'_i,C'_i,D'_i\}}||\hat{\sigma}_h(\{A'_i, B'_i, C'_i, D'_i\}) - \sigma_h||_{\mathbf{K}^{-1}|\Omega_2} \quad (58)$$

In practice, we only keep the first order term $\beta = -1/2$ ($i = 1$) so that only 4 constants need to be determined. At the end of this step, in order to ensure the continuity of the stress vector at the interface Γ_{12} between Ω_1 and Ω_2, we solve over Ω_1 a new problem with Neumann boundary conditions $\overline{\mathbf{F}}_d = \hat{\sigma}_{h|\Omega_2}\mathbf{n}$ over Γ_{12}.

In zone Ω_1, the construction of $\hat{\sigma}_h$ is performed by extending to XFEM the classical procedure [4] based on a prolongation condition and using FE properties of σ_h (see Appendix). In particular, we take into account discontinuities introduced with the enrichment by function $H(\mathbf{x})$.

The classical procedure used in FEM, that consists in constructing equilibrated tractions $\hat{\mathbf{F}}_h$ over the boundary ∂E of each FE element E before solving local elementary problems, can be directly extended to the XFEM framework noticing that XFEM merely consists in adding new basis functions $\{N_i(\mathbf{x})\}$. The prolongation condition for enriched elements thus becomes:

$$\int_E (\hat{\sigma}_h - \sigma_h)\boldsymbol{\nabla} N_i \mathrm{d}E = 0 \quad ; \quad \int_E (\hat{\sigma}_h - \sigma_h)\boldsymbol{\nabla}(N_i H)\mathrm{d}E = 0 \quad (59)$$

and leads to two relations:

$$\begin{aligned} \int_{\partial E} \eta_E \hat{\mathbf{F}}_h N_i \mathrm{d}S &= \int_E (\sigma_h \boldsymbol{\nabla} N_i - \mathbf{f}_d N_i)\mathrm{d}E = \mathbf{Q}_E(i) \\ \int_{\partial E} \eta_E \hat{\mathbf{F}}_h N_i H \mathrm{d}S &= \int_E (\sigma_h \boldsymbol{\nabla}(N_i H) - \mathbf{f}_d N_i H)\mathrm{d}E = \mathbf{Q}^H_E(i) \end{aligned} \quad (60)$$

Writing (60) for all elements connected to node i defines a local problem. This is illustrated in Fig. 8 for a 2D mesh where two of the four quadrangle elements connected to node i are cut by the crack; we thus obtain in this case:

$$\begin{cases} \mathbf{b}_{14} - \mathbf{b}_{21} = \mathbf{Q}_{E_1}(i) \\ \mathbf{b}_{21} - \mathbf{b}_{32} = \mathbf{Q}_{E_2}(i) \\ \mathbf{b}_{32} - \mathbf{b}_{43} = \mathbf{Q}_{E_3}(i) \\ \mathbf{b}_{43} - \mathbf{b}_{14} = \mathbf{Q}_{E_4}(i) \end{cases} \quad \text{and} \quad \begin{cases} \mathbf{b}_{14} - \mathbf{b}^H_{21} = \mathbf{Q}^H_{E_1}(i) \\ \mathbf{b}^H_{21} - \mathbf{b}_{32} = \mathbf{Q}^H_{E_2}(i) \end{cases} \quad (61)$$

with unknowns $\mathbf{b}_{kl} = \int_{\Gamma_{kl}} \eta_{E_k} \hat{\mathbf{F}}_h N_i \mathrm{d}S$ and $\mathbf{b}^H_{kl} = \int_{\Gamma_{kl}} \eta_{E_k} \hat{\mathbf{F}}_h N_i H \mathrm{d}S$. Properties of σ_h imply that $\sum_E \mathbf{Q}_E(i) = \sum_E \mathbf{Q}^H_E(i) = \mathbf{0}$ and ensure that local problems of type (61) are well-posed.

After computing projections $\mathbf{b}_{ij}$ and $\mathbf{b}^H_{ij}$, we construct tractions along element edges. In the case of edges cut by the crack (for instance Γ_{12} in Fig. 8), tractions are searched under the form:

$$\hat{\mathbf{F}}_h = \hat{\mathbf{F}}_i N_i + \hat{\mathbf{F}}_j N_j + \hat{\mathbf{F}}^H_i N_i H + \hat{\mathbf{F}}^H_j N_j H \quad (62)$$

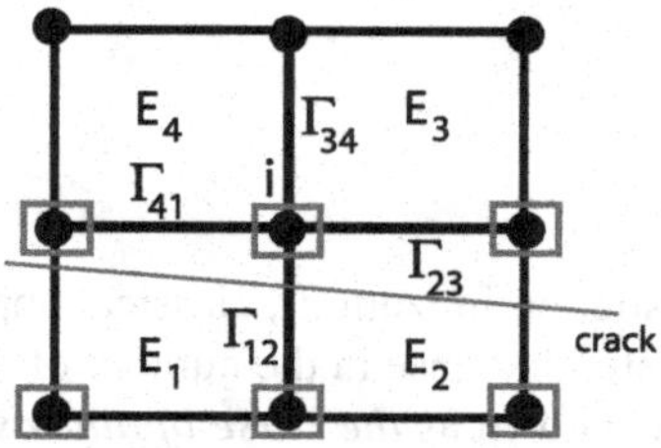

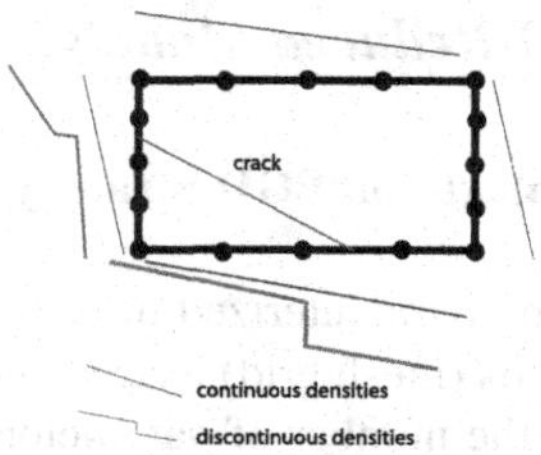

Fig. 8 Patch of elements around node i, with elements E_1 and E_2 cut by the crack, and framed enriched nodes (*left*); local problem to be solved over each element E (*right*)

i.e. as a sum of a continuous part and a discontinuous part. Quantities $\hat{\mathbf{F}}_i$, $\hat{\mathbf{F}}_j$, $\hat{\mathbf{F}}_i^H$, and $\hat{\mathbf{F}}_j^H$ can be easily determined by inverting a small linear system implying previously computed projections. Eventually, an admissible stress field $\hat{\sigma}_{h|E}$ is constructed solving a local problem over each element E with Neumann boundary conditions defined by tractions $\hat{\mathbf{F}}_h$. Here again, an enrichment with HN_i functions is used for elements cut by the crack (Fig. 8).

As an illustration, we consider the cracked structure of Fig. 9 clamped on its bottom side and submitted to a uniform shear force density τ on its top side (mixed mode). Parameters are $\tau = 1$, $E = 210$, $\nu = 0.3$, $L = 16$, and $w = 7$. The structure is discretized using a regular mesh with 1071 Q4 elements.

We consider the quantity of interest K_I. We report in Fig. 9 the values of bounds obtained on this quantity of interest. Quantity $K_{I,corr}$ is a correction term, whereas $K_I^{\pm}$ are upper and lower bounds on the exact value of K_I. The reference value, computed using an overkill mesh, is $K_{I,ref} = 33.93$.

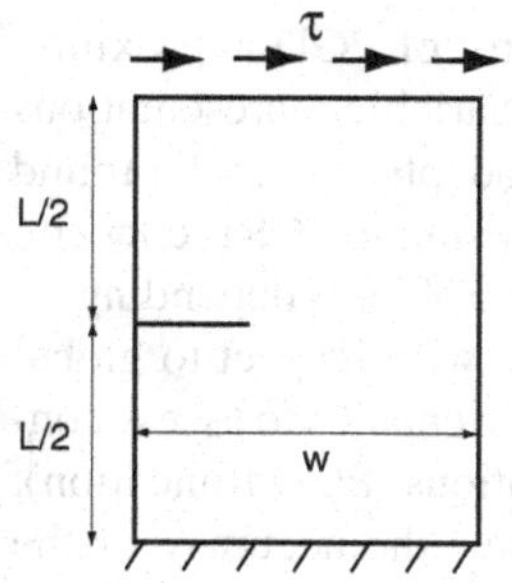

number of elements	$K_{I,h}$	$K_{I,h} + K_{I,corr}$	K_I^-	K_I^+
1071	33.3374	33.6305	32.5714	34.6896
2975	33.3390	33.9100	33.3398	34.4803
5831	33.8579	33.8742	33.5845	34.5845

Fig. 9 Structure with mixed mode loading (*left*), and bounds on K_I for the mixed problem solved with XFEM (*right*)

6.2 PGD-Reduced Models

6.2.1 Context and PGD Strategy

Considering parameterized models (stochastic, optimization), classical approximation methods (mesh grid) lead to an exponential increase in the number of dofs with respect to the number of parameters. This is known as *the curse of dimensionality*, and this rapidly leads to unaffordable computations. Model reduction techniques are alternative tools to address such problems; they have been the object of increasing interest during the previous decade. In particular, an attractive technique called *Proper Generalized Decomposition* (PGD) recently emerged and is currently the topic of various research works [49]. The technique, which leans on ideas initially developed in [50] to solve nonlinear time-dependent problems, is based on variable separation in a spectral approach. The main assets are that no information on the solution is required (contrary to POD) and that the computational cost increases linearly with respect to the number of parameters. PGD basis functions (or modes) are computed on the fly, once for all and in an *offline* process, solving a set of mono-parameter problems with classical techniques. The obtained PGD approximation, that explicitly depends of all model parameters, can then be used in an *online* process for analysis or optimization.

Performances of PGD have been shown in many applications exhibiting changes in loading, boundary or initial conditions, material parameters, geometry, ...taken into account by means of additional coordinates in the model [49]. However, a main difficulty for the transfer and intensive use of PGD reduced models in industry is the control of their reliability. Indeed, certifying the PGD solution is fundamental to perform robust design. The control of the PGD solution requires to master the number of modes which are computed (truncation), but also the numerical methods which are employed in the computation of modes.

There are currently very few works addressing the control of PGD approximations. Basic results on a priori error estimation for separated variable representations are given in [50], whereas a pioneering work dedicated to adaptivity can be found in [51]. A first robust approach for the verification of PGD, using the CRE concept, was proposed in [30]. It applies to linear elliptic or parabolic problems depending on parameters, and provides for certified PGD reduced models with respect to global error or error on outputs of interest. Furthermore, the approach enables to assess contributions of the various error sources (space/time discretizations, PGD truncation), which constitutes relevant information to effectively improve the accuracy of the PGD solution. Performances were shown in [30, 52–54] on several numerical experiments implying a transient thermal model with fluctuating material parameters; we provide in the following basic ideas on the approach.

We consider a transient thermal problem defined on domain $\Omega \subset R^d$ ($d = 1, 2, 3$), with boundary $\partial\Omega$, over the time interval $\mathcal{I} = [0, T]$. A zero temperature is prescribed on boundary $\partial_1\Omega \neq \emptyset$ of $\partial\Omega$ and the time-dependent thermal loading consists of: (i) a given thermal flux $r_d(\mathbf{x}, t)$ on the complementary boundary

$\partial_2\Omega \subset \partial\Omega$; (ii) a source term $f_d(\mathbf{x},t)$ in Ω. Initial boundary conditions are homogeneous. The material that composes Ω is supposed to be heterogeneous and partially unknown. Therefore, the diffusion tensor $\mathbf{K}$ and the thermal capacity c depend on space variable $\mathbf{x}$ but also on a set of N parameters $\mathbf{p} = [p_1, p_2, \dots, p_N]$ belonging to a given bounded domain $\Theta = \Theta_1 \times \Theta_2 \times \dots \times \Theta_N$.

We denote by $(u, \boldsymbol{\varphi})$ the temperature/flux solution of the problem. Defining $\mathcal{U}_{ad} = H_0^1(\Omega) = \{v \in H^1(\Omega), v_{|\partial_1\Omega} = 0\}$, the weak formulation in space of the problem reads for all $(t, \mathbf{p}) \in \mathcal{I} \times \Theta$:

$$\text{Find } u(\mathbf{x},t,\mathbf{p}) \in \mathcal{U}_{ad} \text{ such that } b(u,v) = l(v) \quad \forall v \in \mathcal{U}_{ad} \tag{63}$$

with $u_{|t=0^+} = 0$. The bilinear form $b(\bullet,\bullet)$ and linear form $l(\bullet)$ are defined as:

$$b(u,v) = \int_\Omega \left\{ c\frac{\partial u}{\partial t} v + \mathbf{K}\nabla u \cdot \nabla v \right\} \mathrm{d}\Omega \quad ; \quad l(v) = \int_\Omega f_d v \mathrm{d}\Omega - \int_{\partial_2\Omega} r_d v \mathrm{d}S \tag{64}$$

We now introduce functional spaces $\mathcal{T} = L^2(\mathcal{I})$, $\mathcal{P}_i = L^2(\Theta_i)$, and $L^2(\mathcal{I},\Theta;\mathcal{U}_{ad}) = \mathcal{U}_{ad} \otimes \mathcal{T} \otimes_{n=1}^N \mathcal{P}_n$. The full weak formulation of the problem consists in searching $u \in L^2(\mathcal{I},\Theta;\mathcal{U}_{ad})$, with $\frac{\partial u}{\partial t} \in L^2(\mathcal{I},\Theta;L^2(\Omega))$, such that:

$$B(u,v) = L(v) \quad \forall v \in L^2(\mathcal{I},\Theta;\mathcal{U}_{ad}) \tag{65}$$

with

$$B(u,v) = \int_\Theta \left[\int_{\mathcal{I}} b(u,v)\mathrm{d}t + \int_\Omega c u(\mathbf{x},0^+) v(\mathbf{x},0^+) \mathrm{d}\Omega \right] \mathrm{d}\mathbf{p} \quad ; \quad L(v) = \int_\Theta \int_{\mathcal{I}} l(v) \mathrm{d}t \mathrm{d}\mathbf{p} \tag{66}$$

The approximate solution of (65), from the FEM in space associated to a given time integration scheme and a given grid in the parameter space Θ, can be very costly when the number of parameters increases.

The alternative PGD technique consists in constructing a priori a separated variable representation of the solution u of (65). The approximate PGD solution is searched under the form:

$$u(\mathbf{x},t,\mathbf{p}) \approx u_m(\mathbf{x},t,\mathbf{p}) \equiv \sum_{i=1}^m \psi_i(\mathbf{x}) \lambda_i(t) \Gamma_i(\mathbf{p}) \quad \text{with } \Gamma_i(\mathbf{p}) = \prod_{n=1}^N \gamma_{i,n}(p_n) \tag{67}$$

m is the order (i.e. the number of modes) of the representation, while space functions $\psi_i(\mathbf{x})$, time functions $\lambda_i(t)$, and parameter functions $\gamma_{i,n}(p_n)$ respectively belong to $\mathcal{U}_{ad}$, $\mathcal{T}$ and $\mathcal{P}_n$.

The construction of modes does not require any particular knowledge on u, it is performed on the fly when solving. We give here a classical version of this construction, called *progressive Galerkin* and inspired from classical fixed point algorithms.

We suppose an order $s-1$ PGD approximation has been computed. The order s PGD approximation is then defined as:

$$u_s(\mathbf{x}, t, \mathbf{p}) = u_{s-1}(\mathbf{x}, t, \mathbf{p}) + \psi(\mathbf{x})\lambda(t)\Gamma(\mathbf{p}) \quad \text{with } \Gamma(\mathbf{p}) = \prod_{n=1}^{N} \gamma_n(p_n) \tag{68}$$

ψ, λ, and γ_n ($n = 1, \ldots, N$]) are a priori unknown functions respectively belonging to discretized subspaces $\mathcal{U}_{ad,h}$, $\mathcal{T}_h$, and $\mathcal{P}_{nh}$; we assume that $\mathcal{U}_{ad,h}$ and $\mathcal{T}_h$ verify kinematic constraints and initial conditions, respectively. Starting from initialization $\psi^{(0)}(\mathbf{x})\lambda^{(0)}(t)\Gamma^{(0)}(\mathbf{p})$ for mode s, we construct a new modal representation $\psi^{(1)}(\mathbf{x})\lambda^{(1)}(t)\Gamma^{(1)}(\mathbf{p})$ by means of a Galerkin approach that leads to the following sub-iteration:

→ find $\lambda^{(1)} \in \mathcal{T}_h$ such that:

$$B(u_{s-1} + \psi^{(0)}\lambda^{(1)}\Gamma^{(0)}, \psi^{(0)}\lambda^{*}\Gamma^{(0)}) = L(\psi^{(0)}\lambda^{*}\Gamma^{(0)}) \quad \forall \lambda^{*} \in \mathcal{T}_h \tag{69}$$

→ for $n_0 = 1, \ldots, N$, find $\gamma_{n_0}^{(1)} \in \mathcal{P}_{n_0 h}$ such that:

$$B(u_{s-1} + \psi^{(0)}\lambda^{(1)}\gamma_{n_0}^{(1)}\Gamma_{/n_0}^{(1,0)}, \psi^{(0)}\lambda^{(1)}\gamma^{*}\Gamma_{/n_0}^{(1,0)}) = L(\psi^{(0)}\lambda^{(1)}\gamma^{*}\Gamma_{/n_0}^{(1,0)}) \quad \forall \gamma^{*} \in \mathcal{P}_{n_0 h} \tag{70}$$

with $\Gamma_{/n_0}^{(1,0)} = \prod_{n=1}^{n_0-1} \gamma_n^{(1)} \times \prod_{n=n_0+1}^{N} \gamma_n^{(0)}$;

→ find $\psi^{(1)} \in \mathcal{U}_{ad,h}$ such that:

$$B(u_{s-1} + \psi^{(1)}\lambda^{(1)}\Gamma^{(1)}, \psi^{*}\lambda^{(1)}\Gamma^{(1)}) = L(\psi^{*}\lambda^{(1)}\Gamma^{(1)}) \quad \forall \psi^{*} \in \mathcal{U}_{ad,h} \tag{71}$$

The sub-iteration consists in solving a set of simple problems: the ODE in time coming from (69) is solved with an explicit Euler scheme (time step Δt), the space problem coming from (71) is solved with the FEM (mesh size h), while the solution to problems coming from (70) is explicit.

A few sub-iterations are performed in practice. Moreover, the time function $\lambda^{(j)}(t)$ and parameter functions $\gamma_n^{(j)}(p_n)$ are normalized at each sub-iteration. Let us note that a sub-iteration terminates with space problem (71), which is fundamental for the error estimation technique shown in the following. Optimizations, such as updating of the set $\{\lambda_i\}$ (resp. $\{\Gamma_i\}$) of time functions (resp. parameter functions) or orthogonalization of space modes, are possible.

6.2.2 Construction of Equilibrated Fields and Error Estimation

The proposed verification strategy uses the CRE concept. Let $(\hat{u}, \hat{\varphi})$ be an admissible solution to the problem i.e. verifying (in addition to initial conditions) kinematic constraints and balance equations for all $(t, \mathbf{p}) \in \mathcal{I} \times \Theta$. The CRE measure in the space-time domain, that depends on $\mathbf{p}$, thus reads:

$$E^2_{CRE}(\mathbf{p}) = \frac{1}{2}\int_{\mathcal{I}}\int_{\Omega} \mathbf{K}^{-1}[\hat{\boldsymbol{\varphi}} - \mathbf{K}\nabla\hat{u}]\cdot[\hat{\boldsymbol{\varphi}} - \mathbf{K}\nabla\hat{u}]\mathrm{d}\Omega\mathrm{d}t \equiv \frac{1}{2}|||\hat{\boldsymbol{\varphi}} - \mathbf{K}\nabla\hat{u}|||^2_{\mathbf{K}^{-1}} \tag{72}$$

with $|||\bullet|||_{\mathbf{K}^{-1}}$ the energy norm in the space-time domain, and the extension of the Prager-Synge theorem reads:

$$|||\boldsymbol{\varphi} - \hat{\boldsymbol{\varphi}}_m|||^2_{\mathbf{K}^{-1}} + \frac{1}{2}\int_{\Omega} c(u-\hat{u})^2_{|T}\mathrm{d}\Omega = \frac{1}{2}E^2_{CRE} \quad ; \quad \hat{\boldsymbol{\varphi}}_m = \frac{1}{2}[\hat{\boldsymbol{\varphi}} + \mathbf{K}\nabla\hat{u}] \tag{73}$$

Again, the technical point is the construction of an admissible solution; we explain here how such as solution, denoted $(\hat{u}_m, \hat{\boldsymbol{\varphi}}_m)$, can be obtained with a post-processing of all available information from the computation of the PGD solution u_m.

Constructing an admissible kinematic field $\hat{u}_m(\mathbf{x}, t, \mathbf{p})$ is simple, and we choose it equal to $u_m(\mathbf{x}, t, \mathbf{p})$. Obtaining $\hat{\boldsymbol{\varphi}}_m(\mathbf{x}, t, \mathbf{p})$ is more technical; in order to use classical tools that enable to compute equilibrated tractions (in particular the prolongation condition), we should first build a field $\boldsymbol{\varphi}_m(\mathbf{x}, t, \mathbf{p})$ that satisfies the FE equilibrium for all $(t, \mathbf{p}) \in \mathcal{I} \times \Theta$:

$$\int_{\Omega} \boldsymbol{\varphi}_m \cdot \nabla v\mathrm{d}\Omega = \int_{\Omega}(f_d - c\frac{\partial \hat{u}_m}{\partial t})v\mathrm{d}\Omega - \int_{\partial_2\Omega} r_d v\mathrm{d}S \quad \forall v \in \mathcal{U}_{ad,h} \tag{74}$$

We first assume that the external loading can be written under the radial form:

$$(f_d(\mathbf{x}, t), r_d(\mathbf{x}, t)) = \sum_{j=1}^{J} \alpha_j(t)\left(f_d^j(\mathbf{x}), r_d^j(\mathbf{x})\right) \tag{75}$$

We then compute, for each couple (f_d^j, r_d^j), a field $\boldsymbol{\varphi}_d^j(\mathbf{x})$ verifying the FE equilibrium:

$$\int_{\Omega} \boldsymbol{\varphi}_d^j \cdot \nabla v\mathrm{d}\Omega = \int_{\Omega} f_d^j v\mathrm{d}\Omega - \int_{\partial_2\Omega} r_d^j v\mathrm{d}S \quad \forall v \in \mathcal{U}_{ad,h} \tag{76}$$

This computation is in practice performed with the FEM in displacement. This yields, introducing $\boldsymbol{\varphi}_d = \sum_{j=1}^{J}\alpha_j(t)\boldsymbol{\varphi}_d^j(\mathbf{x})$ in (74), that $\boldsymbol{\varphi}_m$ should verify for all $(t, \mathbf{p}) \in \mathcal{I} \times \Theta$:

$$\int_{\Omega}(\boldsymbol{\varphi}_m - \boldsymbol{\varphi}_d)\cdot\nabla v\mathrm{d}\Omega = -\int_{\Omega} c\frac{\partial \hat{u}_m}{\partial t}v\mathrm{d}\Omega = -\sum_{i=1}^{m} c\dot{\lambda}_i\Gamma_i\int_{\Omega}\psi_i v\mathrm{d}\Omega \quad \forall v \in \mathcal{U}_{ad,h} \tag{77}$$

On the other hand, at the end of sub-iterations to compute each PGD mode $m_0 \in [1, m]$, the condition (71) gives:

$$B(u_{m_0}, \psi^*\lambda_{m_0}\Gamma_{m_0}) = L(\psi^*\lambda_{m_0}\Gamma_{m_0}) \quad \forall\psi^* \in \mathcal{U}_{ad,h} \tag{78}$$

This last relation can be recast under the form:

$$\int_{\Omega} \mathbf{H}_{m_0} \cdot \nabla \psi^* \mathrm{d}\Omega = \int_{\Omega} \sum_{i=1}^{m_0} \left[G_{m_0,i} \psi_i \right] \psi^* \mathrm{d}\Omega \quad \forall \psi^* \in \mathscr{U}_{ad,h} \tag{79}$$

with

$$\mathbf{H}_{m_0} \equiv \int_{\Theta} \int_{\mathscr{I}} \lambda_{m_0} \Gamma_{m_0} (\boldsymbol{\varphi}_d - \mathbf{K} \nabla u_{m_0}) \mathrm{d}t \mathrm{d}\mathbf{p} \quad ; \quad G_{m_0,i} \equiv \int_{\Theta} \int_{\mathscr{I}} c \lambda_{m_0} \Gamma_{m_0} \dot{\lambda}_i \Gamma_i \mathrm{d}t \mathrm{d}\mathbf{p} \tag{80}$$

Consequently, for all $m_0 \in [1, m]$, the term $\mathbf{H}_{m_0}$ equilibrates $\sum_{i=1}^{m_0} G_{m_0,i} \psi_i$ in the FE sense. A simple inversion thus generates terms of the form $\sum_{j=1}^{m} R_{ij} \mathbf{H}_j$ that equilibrate each function ψ_i $(i = 1, \ldots, m)$ in the FE sense. A field $\boldsymbol{\varphi}_m$ that satisfies the FE equilibrium (74) (or (77)) thus reads:

$$\boldsymbol{\varphi}_m = \boldsymbol{\varphi}_d - \sum_{i=1}^{m} \sum_{j=1}^{m} c \dot{\lambda}_i \Gamma_i R_{ij} \mathbf{H}_j \tag{81}$$

From $\boldsymbol{\varphi}_m$, usual techniques can then be used to compute a flux $\hat{\boldsymbol{\varphi}}_m$ that verifies strict equilibrium:

$$\int_{\Omega} \hat{\boldsymbol{\varphi}}_m \cdot \nabla v \mathrm{d}\Omega = \int_{\Omega} (f_d - c \frac{\partial \hat{u}_m}{\partial t}) v \mathrm{d}\Omega - \int_{\partial_2 \Omega} r_d v \mathrm{d}S \quad \forall v \in \mathscr{U}_{ad} \tag{82}$$

This flux reads $\hat{\boldsymbol{\varphi}}_m = \hat{\boldsymbol{\varphi}}_d - \sum_{i=1}^{m} \sum_{j=1}^{m} c \dot{\lambda}_i \Gamma_i R_{ij} \hat{\mathbf{H}}_j$ where $\hat{\boldsymbol{\varphi}}_d$ and $\hat{\mathbf{H}}_j$ are computed solving local problems on each element.

Remark 8 In the case of a stationary problem, terms $\mathbf{H}_{m_0} = \int_{\Theta} \Gamma_{m_0} (\boldsymbol{\varphi}_d - \mathbf{K} \nabla u_{m_0}) \mathrm{d}\mathbf{p}$ $(m_0 = 1, \ldots, m)$ are self-equilibrated in the FE sense. Therefore, $\boldsymbol{\varphi}_m$ and $\hat{\boldsymbol{\varphi}}_m$ can be defined as:

$$\boldsymbol{\varphi}_m(\mathbf{x}, \mathbf{p}) = \boldsymbol{\varphi}_d(\mathbf{x}) + \sum_{m_0=1}^{m} \beta_{m_0}(\mathbf{p}) \mathbf{H}_{m_0}(\mathbf{x}) \quad ; \quad \hat{\boldsymbol{\varphi}}_m(\mathbf{x}, \mathbf{p}) = \hat{\boldsymbol{\varphi}}_d(\mathbf{x}) + \sum_{m_0=1}^{m} \beta_{m_0}(\mathbf{p}) \hat{\mathbf{H}}_{m_0}(\mathbf{x}) \tag{83}$$

where coefficients β_{m_0}, functions of $\mathbf{p}$, are explicitly obtained minimizing $\int_{\Theta} E_{CRE}^2(\mathbf{p}) \mathrm{d}\mathbf{p}$.

From the global error estimate $E_{CRE}^2(\mathbf{p})$ previously defined, it is then easy to construct a local error estimate from adjoint-based techniques. Let $Q(\mathbf{p})$ be a quantity of interest defined by extractors $(\tilde{\boldsymbol{\varphi}}_{\Sigma}, \tilde{f}_{\Sigma})$:

$$Q(\mathbf{p}) = \int_{\mathscr{I}} \int_{\Omega} \left(\nabla u(\mathbf{p}) \cdot \tilde{\boldsymbol{\varphi}}_{\Sigma} + u(\mathbf{p}) \tilde{f}_{\Sigma} \right) \mathrm{d}\Omega \mathrm{d}t \tag{84}$$

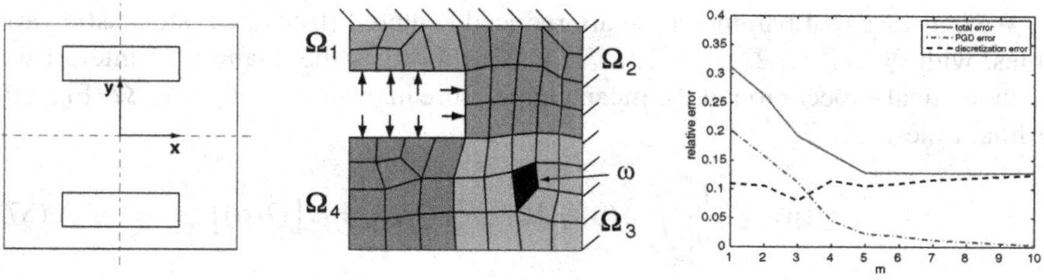

Fig. 10 Representation of the multi-parameter 2D problem (*left*), and relative local error estimate and indicators on the various error sources with respect to the number m of PGD modes for Q_Θ (*right*)

We introduce the associated adjoint problem, and we compute an approximate (resp. admissible) PGD solution $(\tilde{u}_{\tilde{m}}, \tilde{\varphi}_{\tilde{m}})$ (resp. $(\hat{\tilde{u}}_{\tilde{m}}, \hat{\tilde{\varphi}}_{\tilde{m}})$) for this problem. In practice, the PGD solution to the adjoint problem is performed with an order $\tilde{m}$ potentially different from m, and introducing enrichment functions locally. We obtain the bounding:

$$|Q(\mathbf{p}) - Q_m(\mathbf{p}) - Q_{corr}(\mathbf{p})| \leq E_{CRE}(\mathbf{p}).\tilde{E}_{CRE}(\mathbf{p}) \tag{85}$$

where $Q_{corr}(\mathbf{p})$ is a computable correction term, and $E_{CRE}(\mathbf{p})$ (resp. $\tilde{E}_{CRE}(\mathbf{p})$) is the constitutive relation error for the reference problem (resp. adjoint problem). Therefore, guaranteed bounds on the local error $Q(\mathbf{p}) - Q_m(\mathbf{p})$ (or directly on $Q(\mathbf{p})$) can be obtained for any value $\mathbf{p}$ of the model parameters.

As an example, we consider the 2D structure represented on Fig. 10. It is a cross section Ω with two rectangular holes in which a fluid circulates. Using symmetries, only a quarter of the structure is considered. A flux $r_d(t) = -1$ is applied on hole boundaries whereas a source term $f_d(x, y) = 200xy$ is applied in Ω. A zero temperature is imposed on the remainder of $\partial\Omega$. We consider that the diffusion coefficient $\mathbf{K}$ (isotropic behavior) fluctuates but remains piecewise homogeneous, i.e. homogeneous in each of the four subdomains Ω_i $(i = 1, 2, 3, 4)$ defined on Fig. 10 and such that $\Omega_i \cap \Omega_j = \emptyset$, $\overline{\Omega_1 \cup \Omega_2 \cup \Omega_3 \cup \Omega_4} = \Omega$. The thermal capacity c is assumed to be homogeneous in the whole domain Ω. These two material coefficients are thus defined by 5 parameters $(\theta_1, \theta_2, \theta_3, \theta_4, \theta_5)$ such that:

$$K(\mathbf{x}, \theta_i) = 1 + \sum_{i=1}^{4} g_i I_{\Omega_i}(\mathbf{x})\theta_i \qquad c(\mathbf{x}, \theta_5) = 1 + 0.2\,\theta_5 \tag{86}$$

with $[g_1, g_2, g_3, g_4] = [0.1; 0.1; 0.2; 0.05]$, and $I_{\Omega_i}(\mathbf{x})$ referring to the indicatrix function of subdomain Ω_i.

The resulting solution $u(\mathbf{x}, t, \mathbf{p})$, with $\mathbf{p} = [\theta_1, \theta_2, \theta_3, \theta_4, \theta_5]$, is searched using the PGD technique; with an initial discretization made of 50 Q4 elements in space and 1000 time steps.

We consider that parameters θ_i are reduced centered (truncated) stochastic variables, with $\theta_i \in [-2, 2]$ $(i = 1, \dots, 5)$. We choose as the quantity of interest the mathematical expectation of the mean temperature in the local zone $\omega \subset \Omega$ (Fig. 10) at final time T:

$$Q(\mathbf{p}) = \frac{1}{|\omega|} \int_{\omega} u(\mathbf{p})_{|T} \mathrm{d}\Omega \quad ; \quad Q_{\Theta} = \mathrm{E}_{\Theta}\left[Q(\mathbf{p})\right] \tag{87}$$

The relative local error $\int_{\Theta} E_{CRE}.\tilde{E}_{CRE}\mathrm{d}\mathbf{p}/(Q_{\Theta,m} + Q_{\Theta,corr})$ associated to Q_{Θ}, as well as relative indicators on discretization and truncation errors, are given in Fig. 10 with respect to m.

6.2.3 Adaptive Strategy

The error on Q comes from two sources: (i) the order m truncation of the PGD representation (67); (ii) the space-time discretization used to compute modes numerically. We can write:

$$Q - Q_m = [Q(u) - Q(u^{h,\Delta t})] + [Q(u^{h,\Delta t}) - Q(u_m)] = \Delta Q_{dis} + \Delta Q_{trunc} \tag{88}$$

with ΔQ_{trunc} (resp. ΔQ_{dis}) the part on the error on Q coming from the PGD truncation (resp. from the discretization). $u^{h,\Delta t}$ is the solution to the problem arising from the discretization of (65) and seen as the reference problem to define the error ΔQ_{trunc}.

To effectively control the process of PGD computation, we introduce an error indicator for each error source. The evaluation of ΔQ_{trunc} is performed from CRE taking as the reference problem the discretized problem (with solution $u^{h,\Delta t}$) which can be put under the form:

$$\mathbf{U}_h^1 = \mathbf{0} \quad ; \quad \mathbb{M}(\mathbf{p})\frac{\mathbf{U}_h^{k+1} - \mathbf{U}_h^k}{\Delta t} + \mathbb{K}(\mathbf{p})\mathbf{U}_h^k = \mathbf{F}_h^k \quad \forall k \geq 1 \tag{89}$$

with $\mathbf{U}_h^k$ the vector of nodal unknowns of u at time point k. An admissible solution is reconstructed in the sense of the discretized problem by direct post-processing of available information. The evaluation of ΔQ_{dis} is then obtained by the difference between those of $Q - Q_m$ and of ΔQ_{trunc}.

From then on, a simple adaptive strategy consists in evaluating, after the computation of each PGD mode, the various error sources on Q and to adapt with respect to the dominating source: if $|\Delta Q_{dis}|$ is dominating, we define a finer discretization (up to obtaining $|\Delta Q_{dis}| < |\Delta Q_{trunc}|$); if $|\Delta Q_{trunc}|$ is dominating, we compute the next PGD mode without modifying the discretization parameters.

For the same test case, we perform the adaptive procedure (greedy algorithm) coming from the evaluation of error sources. The convergence of the relative local error obtained for Q_{Θ} is shown in Fig. 11. Vertical evolutions of the curves indicate

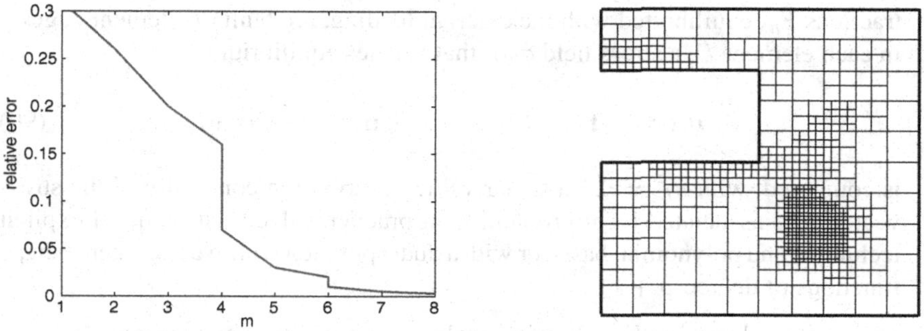

Fig. 11 Convergence of the relative error with respect to the number m of PGD modes in the adaptive strategy for Q_Θ (*left*), and refined mesh used for the computation of PGD mode $m = 8$ after applying the adaptive strategy (*right*)

mesh refinement. We observe that the procedure enables to effectively decrease the error on Q_Θ. Up to mode 4, the initial coarse discretization is sufficient to compute the PGD solution, and the local error decreases when computing a new mode. After $m = 4$, PGD modes represent finer details that require the modification of the discretization parameters. This is illustrated in Fig. 11 where we represent the refined mesh (*quad-tree* structure) used for the computation of the eighth PGD mode, for the adaptive strategy related to Q_Θ.

7 Conclusions and Prospects

We have presented the main aspects and capabilities of error estimation techniques based on the CRE concept. These robust techniques provide for bounds which are both guaranteed and accurate for a large class of mechanical problems, as well as relevant information to drive adaptive processes. Several remaining research issues will be addressed in the near future, such as accurate bounds for complex nonlinear material behavior (with possible softening, instabilities, or large deformations).

Appendix: Construction of Equilibrated Stress Fields

The construction of a statically admissible field is a key point of error estimation methods based on CRE. It particularly enables to obtain guaranteed error bounds for a large set of mechanical problems. A general construction approach, based on a post-processing of the FE stress field σ_h, has been introduced in [4, 35]. This approach, recently named EET (Element Equilibration Technique), can be decomposed in two steps:

1. tractions $\hat{\mathbf{F}}_h$, equilibrated with the external loading, are built on element edges;
2. in each element E, a stress field $\hat{\sigma}_{h|E}$ that verifies equilibrium:

$$\mathbf{div}\,\hat{\sigma}_h + \mathbf{f}_d = \mathbf{0} \text{ in } E \quad ; \quad \hat{\sigma}_h \mathbf{n} = \eta_E \hat{\mathbf{F}}_h \text{ on } \partial E \tag{90}$$

is computed, with $\eta_E = \pm 1$ a scalar value ensuring the continuity of the stress vector. The associated local problem is in practice solved with a quasi-explicit technique and polynomial basis, or with a dual approach with p enrichment (shape functions of degree $p + k$).

The first step leans on the following prolongation (energy) condition:

$$\int_E (\hat{\sigma}_h - \sigma_h)\nabla\phi_i \mathrm{d}E = 0 \quad \Longrightarrow \quad \int_{\partial E} \hat{\sigma}_h \cdot \mathbf{n}\phi_i \mathrm{d}S = \int_E (\sigma_h \nabla\phi_i - \mathbf{f}_d \cdot \phi_i)\mathrm{d}E \quad \forall i \tag{91}$$

where ϕ_i is the FE shape function associated to node i. This condition, which ensures equilibration of $\hat{\mathbf{F}}_h$ over E (as $\sum_i \phi_{i|E} = 1$), leads to the solution to a system of the form:

$$\sum_{r=1}^{R_n} \mathbf{b}_n^r(i) = \mathbf{Q}_{E_n}(i) \quad \forall n = 1, \ldots, N \tag{92}$$

over the set of N elements connected to each node i. R_n is the number of edges for element E_n connected to node i, $\mathbf{Q}_{E_n}(i) = \int_{E_n} (\sigma_h \nabla\phi_i - \mathbf{f}_d \phi_i)\mathrm{d}E$, and unknowns $\mathbf{b}_n^r(i)$ are projections of tractions defined as $\hat{\mathbf{b}}_n^r(i) = \int_{\Gamma_{E_n}^r} \eta_{E_n} \hat{\mathbf{F}}_h \phi_i \mathrm{d}S$. Existence of a solution for each system is ensured by the equilibrium property (in the FE sense) verified by σ_h, and uniqueness may be obtained minimizing a cost function.

In [26], a new hybrid method called EESPT (Element Equilibration + Star Patch Technique) was introduced for the construction of admissible stress fields. As an intermediary between EET and SPET (flux-free [55, 56]) methods, it enables a nice compromise between accuracy of the computed stress fields, computational cost, and practical implementation in engineering softwares. The EESPT method still has two steps and leans on the construction of equilibrated tractions $\hat{\mathbf{F}}_h$ on element edges. The main change is in the way the tractions are constructed, with an increasing flexibility brought by a Partition of Unity Method (PUM); this leads to patch problems solved in an automatic and non-intrusive manner, from classical FE tools. The computation of $\hat{\sigma}_h$, over each element and from tractions $\hat{\mathbf{F}}_h$, remains unchanged and can be parallelized.

A comparison between EET, SPET and EESPT methods was performed in [36] on several industrial applications, one of them being the structure presented in Fig. 2. It was observed that the SPET method is more accurate than EET and EESPT methods, but it requires higher computational cost. The EESPT method, which provides results comparable to those of the EET method, seems to be a nice compromise between accuracy, computational cost and implementation issues.

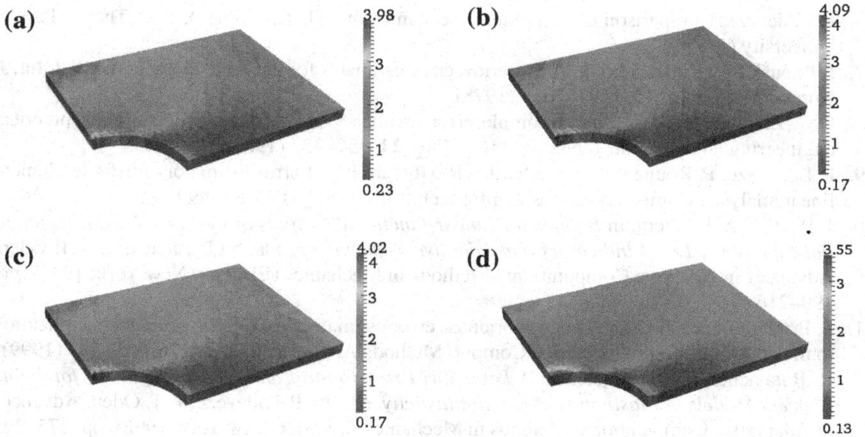

Fig. 12 Norm of the FE stress field (**a**), and admissible stress fields computed with EET (**b**), standard SPET (**c**), and enhanced SPET (**d**) methods

In [57], an improved version of the EESPT method was studied. It uses ideas developed in [58] by considering a weak prolongation condition applied to high degree shape functions (non-vertex nodes). This results in a local minimization of the complementary energy and leads to optimized tractions in selected regions, particularly those with distorted elements or high gradients. The improved version of the EESPT method having a higher computational cost, criteria were introduced to select zones in which this version should be employed to get a nice compromise between accuracy and cost. One example is that of a plate with a hole subjected to a unit traction force (see Fig. 12). EET, standard EESPT, and enhanced EESPT methods were used to compute, from σ_h, a SA stress field $\hat{\sigma}_h$ and derive the associated CRE error estimate. Two criteria were introduced to detect zones in which the enhanced SPET method should be used; the first criterion is based on the element shape (distorsion level) and thus relies on the local quality of the mesh, whereas the second criterion considers local error contributions.

References

1. I. Babuška, T. Strouboulis, *The Finite Element Method and Its Reliability* (Oxford University Press, Oxford, 2001)
2. N.E. Wiberg, P. Diez, Special issue. Comput. Methods Appl. Mech. Eng. **195**, 4–6 (2006)
3. P. Ladevèze, J.-T. Oden, *Advances in Adaptative Computational Methods in Mechanics* (Elsevier, New York, 1998)
4. P. Ladevèze, J.-P. Pelle, *Mastering Calculations in Linear and Nonlinear Mechanics* (Springer, New York, 2004)
5. E. Stein, *Error Controlled Adaptive Finite Elements in Solid Mechanics* (Wiley, Chichester, 2003)

6. P. Ladevèze, Comparison of continuum mechanics models (in French). PhD Thesis, Paris 6 University (1975)
7. I. Babuška, W.C. Rheinboldt, A posteriori error estimates for the finite element method. Int. J. Numer. Meth. Eng. **12**, 1597–1615 (1978)
8. O.C. Zienkiewicz, J.Z. Zhu, A simple error estimator and adaptive procedure for practical engineering analysis. Int. J. Numer. Meth. Eng. **24**, 337–357 (1987)
9. P. Ladevèze, P. Rougeot, P. Blanchard, J.P. Moreau, Local error estimators for finite element linear analysis. Comput. Methods Appl. Mech. Eng. **176**, 231–246 (1999)
10. J. Peraire, A.T. Patera, in *Bounds for Linear-functional Outputs of Coercive Partial Differential Equations: Local Indicators and Adaptive Refinements*, eds. by P. Ladevèze, J.-T. Oden. Advances in Adaptive Computational Methods in Mechanics (Elsevier, New York, 1998), pp. 199–216
11. S. Prudhomme, J.T. Oden, On goal-oriented error estimation for elliptic problems: application to the control of pointwise errors. Comput. Methods Appl. Mech. Eng. **176**, 313–331 (1999)
12. R. Rannacher, F.T. Suttmeier, in *A Posteriori Error Control and Mesh Adaptation for Finite Element Models in Elasticity and Elasto-plasticity*, eds. by P. Ladevèze, J.-T. Oden. Advances in Adaptative Computational Methods in Mechanics (Elsevier, New York, 1998), pp. 275–292
13. T. Strouboulis, I. Babuška, D. Datta, K. Copps, S.K. Gangaraj, A posteriori estimation and adaptive control of the error in the quantity of interest—part 1: a posteriori estimation of the error in the Von Mises stress and the stress intensity factors. Comput. Methods Appl. Mech. Eng. **181**, 261–294 (2000)
14. R. Becker, R. Rannacher, A feed-back approach to error control in finite element methods: basic analysis and examples. East-West J. Numer. Math. **4**, 237–264 (1996)
15. K. Eriksson, D. Estep, P. Hansbo, C. Johnson, in *Introduction to Adaptive Methods For Partial Differential Equations*, ed. by A. Isereles. Acta Numerica (Cambridge University Press, Cambridge, 1995), pp. 105–159
16. F. Cirak, E. Ramm, A posteriori error estimation and adaptivity for linear elasticity using the reciprocal theorem. Comput. Methods Appl. Mech. Eng. **156**, 351–362 (1998)
17. N. Parès, J. Bonet, A. Huerta, J. Peraire, The computation of bounds for linear-functional outputs of weak solutions to the two-dimensional elasticity equations. Comput. Methods Appl. Mech. Eng. **195**(4–6), 406–429 (2006)
18. H.J. Greenberg, The determination of upper and lower bounds for the solution of Dirichlet problem. J. Math. Phys. **27**, 161–182 (1948)
19. K. Washizu, Bounds for solutions of boundary value problems in elasticity. J. Math. Phys. **32**, 117–128 (1953)
20. P. Ladevèze, Upper error bounds on calculated outputs of interest for linear and nonlinear structural problems. Comptes Rendus Acadmie des Sciences—Mcanique, Paris **334**, 399–407 (2006)
21. P. Ladevèze, B. Blaysat, E. Florentin, Strict upper bounds of the error in calculated outputs of interest for plasticity problems. Comput. Methods Appl. Mech. Eng. **245–246**, 194–205 (2012)
22. J. Waeytens, L. Chamoin, P. Ladevèze, Guaranteed error bounds on pointwise quantities of interest for transient viscodynamics problems. Comput. Mech. **49**(3), 291–307 (2012)
23. L. Chamoin, P. Ladevèze, Bounds on history-dependent or independent local quantities in viscoelasticity problems solved by approximate methods. Int. J. Numer. Meth. Eng. **71**(12), 1387–1411 (2007)
24. L. Chamoin, P. Ladevèze, A non-intrusive method for the calculation of strict and efficient bounds of calculated outputs of interest in linear viscoelasticity problems. Comput. Methods Appl. Mech. Eng. **197**(9–12), 994–1014 (2008)
25. P. Ladevèze, L. Chamoin, Calculation of strict error bounds for finite element approximations of nonlinear pointwise quantities of interest. Int. J. Numer. Meth. Eng. **84**, 1638–1664 (2010)
26. P. Ladevèze, L. Chamoin, E. Florentin, A new non-intrusive technique for the construction of admissible stress fields in model verification. Comput. Methods Appl. Mech. Eng. **199**(9–12), 766–777 (2010)

27. P. Ladevèze, F. Pled, L. Chamoin, New bounding techniques for goal-oriented error estimation applied to linear problems. Int. J. Numer. Meth. Eng. **93**, 1345–1380 (2013)
28. L. Chamoin, E. Florentin, S. Pavot, V. Visseq, Robust goal-oriented error estimation based on the constitutive relation error for stochastic problems. Comput. Struct. **106–107**, 189–195 (2012)
29. P. Ladevèze, Strict upper error bounds for computed outputs of interest in computational structural mechanics. Comput. Mech. **42**(2), 271–286 (2008)
30. P. Ladevèze, L. Chamoin, On the verification of model reduction methods based on the Proper Generalized Decomposition. Comput. Methods Appl. Mech. Eng. **200**, 2032–2047 (2011)
31. P. Ladevèze, J. Waeytens, Model verification in dynamics through strict upper error bounds. Comput. Methods Appl. Mech. Eng. **198**(21–26), 1775–1784 (2009)
32. J. Panetier, P. Ladevèze, F. Louf, Strict bounds for computed stress intensity factors. Comput. Struct. **871**(15–16), 1015–1021 (2009)
33. J. Panetier, P. Ladevèze, L. Chamoin, Strict and effective bounds in goal-oriented error estimation applied to fracture mechanics problems solved with the XFEM. Int. J. Numer. Meth. Eng. **81**(6), 671–700 (2010)
34. P. Ladevèze, E.A.W. Maunder, A general method for recovering equilibrating element tractions. Comput. Methods Appl. Mech. Eng. **137**, 111–151 (1996)
35. P. Ladevèze, D. Leguillon, Error estimate procedure in the finite element method and application. SIAM J. Numer. Anal. **20**(3), 485–509 (1983)
36. F. Pled, L. Chamoin, P. Ladevèze, On the techniques for constructing admissible stress fields in model verification: performances on engineering examples. Int. J. Numer. Meth. Eng. **88**(5), 409–441 (2011)
37. P. Ladevèze, J.L. Gastine, P. Marin, J.P. Pelle, Accuracy and optimal meshes in finite element computation for nearly incompressible materials. Comput. Methods Appl. Mech. Eng. **94**(3), 303–314 (1992)
38. W. Prager, J.L. Synge, Approximation in elasticity based on the concept of functions spaces. Q. Appl. Math. **5**, 261–269 (1947)
39. P. Ladevèze, N. Moës, A new a posteriori error estimation for nonlinear time-dependent finite element analysis. Comput. Methods Appl. Mech. Eng. **157**, 45–68 (1998)
40. N. Moës, J. Dolbow, T. Belytschko, A finite element method for crack growth without remeshing. Int. J. Numer. Meth. Eng. **46**, 131–150 (1999)
41. T. Strouboulis, I. Babuška, K. Copps, The design and analysis of the generalized finite element method. Comput. Methods Appl. Mech. Eng. **181**(1–3), 43–69 (2000)
42. R.D. Mindlin, Force at a point in the interior of a semi-infinite solid. J. Phys. **7**, 195–202 (1936)
43. S. Vijayakumar, E.C. Cormack, Green's functions for the biharmonic equation: bonded elastic media. SIAM J. Appl. Math. **47**(5) (1987)
44. M. Dahan, M. Predeleanu, Solutions fondamentales pour un milieu élastique à isotropie transverse. ZAMP **31**, 413–424 (1980)
45. P. Ladevèze, E. Florentin, Verification of stochastic models in uncertain environments using the constitutive relation error method. Comput. Methods Appl. Mech. Eng. **196**, 225–234 (2006)
46. L. Gallimard, J. Panetier, Error estimation of stress intensity factors for mixed-mode cracks. Int. J. Numer. Meth. Eng. **68**(3), 299–316 (2006)
47. T.J. Stone, I. Babuška, A numerical method with a posteriori error estimation for determining the path taken by a propagating crack. Comput. Methods Appl. Mech. Eng. **160**, 245–271 (1998)
48. I. Babuška, A. Miller, The post-processing approach in the finite element method—part 2: the calculation of stress intensity factors. Int. J. Numer. Methods Eng. **20**, 1111–1129 (1984)
49. F. Chinesta, A. Ammar, E. Cueto, Recent advances and new challenges in the use of the proper generalized decomposition for solving multidimensional models. Arch. Comput. Methods Eng. **17**(4), 327–350 (2010)
50. P. Ladevèze, *Nonlinear Computational Structural Mechanics—New Approaches and Non-Incremental Methods of Calculation* (Springer, New York, 1999)

51. A. Ammar, F. Chinesta, P. Diez, A. Huerta, An error estimator for separated representations of highly multidimensional models. Comput. Methods Appl. Mech. Eng. **199**, 1872–1880 (2010)
52. P.E. Allier, L. Chamoin, P. Ladevèze, Proper Generalized Decomposition computational methods on a benchmark problem. Submitted to AMSES (2015)
53. L. Chamoin, P. Ladevèze, Robust control of PGD-based numerical simulations. Eur. J. Comput. Mech. **21**(3–6), 195–207 (2012)
54. P. Ladevèze, L. Chamoin, in *Toward Guaranteed PGD-reduced Models*, eds. by G. Zavarise, D.P. Boso. Bytes and Science (CIMNE, 2012)
55. N. Pares, P. Diez, A. Huerta, Subdomain-based flux-free a posteriori error estimators. Comput. Methods Appl. Mech. Eng. **195**(4–6), 297–323 (2006)
56. J.P. Moitinho de Almeida, E.A.W. Maunder, Recovery of equilibrium on star patches using a partition of unity technique. Int. J. Numer. Meth. Eng. **79**, 1493–1516 (2009)
57. F. Pled, L. Chamoin, P. Ladevèze, An enhanced method with local energy minimization for the robust a posteriori construction of equilibrated stress fields in finite element analyses. Comput. Mech. **49**(3), 357–378 (2012)
58. E. Florentin, L. Gallimard, J.P. Pelle, Evaluation of the local quality of stresses in 3D finite element analysis. Comput. Methods Appl. Mech. Eng. **191**, 4441–4457 (2002)

Printed in the United States
By Bookmasters